Beyond the Body Farm

ALSO BY DR. BILL BASS AND JON JEFFERSON

FICTION

Carved in Bone

Flesh and Bone

NONFICTION

Death's Acre

Beyond the Body Farm

A LEGENDARY BONE DETECTIVE EXPLORES MURDERS, MYSTERIES, AND THE REVOLUTION IN FORENSIC SCIENCE

Dr. Bill Bass and Jon Jefferson

An Imprint of HarperCollins*Publishers*

HarperCollins books may be purchased for educational, business, or sales promotional use. For information please write: Special Markets Department, HarperCollins Publishers, 10 East 53rd Street, New York, NY 10022.

FIRST HARPERLUXE EDITION

HarperLuxe™ is a trademark of HarperCollins Publishers.

Library of Congress Cataloging-in-Publication Data is available upon request.

ISBN: 978-0-06-136698-7

ISBN-10: 0-06-136698-6

07 08 09 10 11 WBC/RRD 10 9 8 7 6 5 4 3 2 1

In memory of Dr. Wilton Krogman and Dr. Charlie Snow, pioneering forensic anthropologists and inspiring teachers

CONTENTS

INTRODUCTION: A HALF CENTURY OF FORENSIC EVOLUTION AND REVOLUTION

One April morning fifty-one years ago, Charlie Snow, my anthropology professor at the University of Kentucky, walked into the bone lab as I was hunched over a tray of bones, and asked if I would like to come along on a human identification case. In that one moment, Dr. Snow changed the course of my life and my work forever.

Dr. Snow had been asked by a lawyer whether he could determine the identity of some burned human remains found in a truck outside Lexington—remains found in the charred cab alongside those of the driver. Both the driver and his passenger (believed to be his common-law wife) had been killed when an A&P grocery truck crossed the highway centerline, causing a fiery collision that killed all three people in the vehicles. Dr. Snow told the lawyer that, yes, he could certainly tell whether the burned remains were indeed those of the driver's common-law wife, provided the woman's dental records could be found. It wasn't an idle boast; as director of the U.S. Army's Central Identification Laboratory, Dr. Snow had spent years, during and after World War II, identifying the decomposed, fragmented, and incinerated remains of American soldiers. Snow was in the vanguard of a specialty that decades later would come to be called "forensic anthropology": using the knowledge and techniques of physical anthropology, which traditionally focuses on the study of ancient human bones, to help solve crimes, especially by identifying unknown murder victims and determining how they were killed. Even if it's not possible to identify a victim by name—and I have a shelfful of skeletons where that wasn't possible—a forensic anthropologist can still help by giving the police details about the victim's race, sex, stature, handedness (muscle attachment points tend to be bigger in a right-handed person's right arm, not surprisingly), and manner of death: was the victim stabbed, shot, strangled, bludgeoned, or killed in some other manner that left telltale marks on the bone?

One reason Dr. Snow invited me to accompany him on that life-changing identification case was because I had a car and he did not, though I like to believe the car was far less a factor than my budding brilliance with bones. At any rate, I drove us to the rural church cemetery where the passenger had been buried. The remains inside the waterlogged coffin were slimy and smelly—a far cry from the lustrous ivory-colored bones I'd been studying in the lab the day before. In fact, when the coffin's lid was opened, the sight and the smell were so overpowering that I promptly threw up.

That was fifty-one years and hundreds of forensic cases ago. I'm happy to report that over the intervening years, I have never again thrown up during a forensic case. I'm also happy to report that over those same years, the forensic sciences—the crime-solving versions of anthropology, entomology, odontology (dentistry), genetics, and other sciences that help catch and convict killers—have made quantum leaps, advancing in ways I could not have imagined as I doubled over in that muddy Kentucky cemetery beside Dr. Snow.

I don't mean to imply that forensic science, or forensic scientists, were primitive or backward in the 1950s. After studying with Dr. Snow at Kentucky, I went to the University of Pennsylvania, where I did my Ph.D. studies under Dr. Wilton Krogman, who was internationally renowned as a "bone detective." Krogman wasn't called a forensic anthropologist—that term hadn't been coined yet—but I've never seen anyone better, before or since, at finding clues in human bones: at listening to the secrets the dead can whisper to reveal who they once were and how they were killed. Krogman's particular area of expertise was the skeletal growth and development of children, especially their teeth. For this reason, he tended to have dozens of orthodontists studying under him at any given time. During my years at Penn I was virtually the only forensic anthropologist he taught, and although I didn't have formal training in dentistry or orthodontics, I absorbed a wealth of knowledge about human teeth, especially how they could shed light on a murder victim's age and identity.

One of the most important lessons I've learned during my career is that justice is a team effort. In the course of any given murder case, that team may include uniformed police officers, plainclothes detectives, crime scene and lab technicians, fingerprint experts, medical examiners ("forensic physicians," you might say), firearms and ballistics examiners, toxicologists, forensic dentists, and DNA specialists.

From a broader perspective, though, forensic teamwork extends not just across scientific specialties but across decades of research and innovation. I stand on the shoulders of the towering Dr. Krogman; for his part, Krogman stood on the shoulders of T. Wingate

Todd, a legendary anatomist at Case Western Reserve University in Cleveland, and the first scientist to note that as we age, our pubic bones undergo steady, consistent changes—changes that can indicate the age of an unknown skeleton. Other early giants in adapting the techniques of archaeology and anthropology to shed light on modern murders included Aleš Hrdlička and T. Dale Stewart, physical anthropologists at the Smithsonian Institution. Between the 1930s and the 1960s, Hrdlička and then Stewart consulted on hundreds of cases for the Federal Bureau of Investigation, which was located just a bone's throw away from the Smithsonian. In the process, they—like Krogman—helped define the tools, techniques, and capabilities of forensic anthropology.

In 1972 the Physical Anthropology Section of the American Academy of Forensic Sciences held its first meeting; five years later, a handful of us created the American Board of Forensic Anthropologists. At one point in my career, I'd trained roughly two-thirds of America's board-certified forensic anthropologists; now, since I've retired and other teachers have continued to turn out Ph.D.'s, that percentage is lower. Still, if you look at the "family tree" of forensic anthropology, the limb that hangs down below my name is a gratifyingly big one, branching into scores of respected names—scientists working at institutions as varied as the Smithsonian Institution, the Central Identification Laboratory, the Federal Bureau of Investigation, the Georgia Bureau of Investigation, the Kentucky medical examiner's office, and numerous universities, including (of course!) the University of Tennessee, which has one of the premier forensic anthropology programs in the world.

The centerpiece, or at least the most famous component, of UT's forensic anthropology program is, of course, the Anthropology Research Facility, which is far better known (to the chagrin of some of my younger colleagues) as the Body Farm. I'm often asked how and why I created the Body Farm. I wish I could answer that the vision sprang, full-blown, into my brilliant academic brain, but the fact is, like many scientific journeys, this one unfolded a step or two at a time. From 1960 to 1971, I taught anthropology at the University of Kansas at Lawrence, and during those eleven years, I sometimes identified skeletal remains for Kansas law enforcement officials, ranging from local police to the Kansas Bureau of Investigation; eventually I became friends with Harold Nye, who played a key role in catching the killers profiled in Truman Capote's true-crime classic, *In Cold Blood*, and who later became director of the KBI.

In 1971 I moved to Knoxville to head the Anthropology Department at UT. When I did, the Tennessee state medical examiner, who knew me slightly, asked if I would be willing to serve as the state's forensic anthropologist, helping law enforcement agencies identify bodies. What I didn't foresee when I said yes was the dramatic difference between Tennessee murder victims and Kansas murder victims. In Kansas, when police asked me to help identify someone, they generally brought me a box of dry bones; sometimes there would be bits of mummified tissue clinging to the bone, but for the most part, my cases in Kansas—a big, sparsely populated, relatively dry region—involved skeletal remains. In Tennessee, on the other hand—half as big, twice as many people, and several times as much rainfall—the victims tended to be fresher, smellier, and infinitely buggier, swarming with maggots, the wormlike larvae of blowflies. And when a Tennessee police officer or district attorney would ask me how long one of these corpses had been ripening, I had no solid scientific basis for an answer. So I decided to remedy that gap in my own knowledge. In 1980, on a couple of acres of junk land behind the University of Tennessee Medical Center—a patch of scraggly woods with a charred open area at the center, where the hospital had burned trash for years—I poured a concrete pad measuring sixteen feet square; atop this pad I built a chain-link enclosure, complete with a chain-link "roof." I planned to place, inside the fence and safe from predators (except for predators small enough to slip through the weave of the wire), human bodies, which my graduate students and I would observe closely, recording the sequence and the timing of human decomposition during the extended postmortem interval.

We received our first donated body for study in May of 1981. To preserve the anonymity of the donors, I instituted a numbering system so that research reports would refer to bodies by number, not name. That first body in 1981 became body 1-81; it was soon followed by 2-81, 3-81, and 4-81. In 1982, the numbering sequence would begin with 1-82, 2-82, and so on. (The numbering system for forensic cases is similar to the system for numbering donated bodies, but in forensic cases, the year comes first: our first forensic case of 1981 was 81-1.)

The body donations came slowly at first: we relied heavily on unclaimed bodies from medical examiners around the state. Our numbers didn't even make it out of the single digits during our early years; now, though, so many people know about our research and want to support it that we're getting into the triple digits—well over a hundred donated bodies—every year, with donated bodies surpassing unclaimed bodies by wider and wider margins. One early and influential reason for the rising interest in our research was Patricia Cornwell's novel *The Body Farm,* a blockbuster that shot up the *New York Times* bestseller list in the fall of 1994. Cornwell didn't actually coin the nickname "Body Farm" (that dubious honor, as best we can tell, belongs to an FBI fingerprint expert named Ivan Futrell), but she sure did put us on the map. In the years since her book catapulted us to fame, the Body Farm has been featured in numerous television documentaries, newspaper and magazine articles, radio reports, and, in the past couple of years, a bestselling crime-fiction series, the Body Farm Novels, that Jon Jefferson and I are writing under the pen name "Jefferson Bass." The plots and many of the characters in our novels are fictional, but the science is factual, based on more than a quarter century's worth of experiments at the Anthropology Research Facility. As the Body Farm has grown more famous, it has also grown larger; at the moment it encompasses two or three wooded acres, enclosed within a high wooden fence. With the huge increase in body donations in recent years, however, that's not nearly big enough. Fortunately, the university has said it wants to expand the facility by adding another eleven acres. If business keeps growing at anywhere near the current rate, though, we'll outgrow the addition within a few years. People these days are just dying to get into the Body Farm. . . .

Not surprisingly, when we began our research program back in the early 1980s, our experiments were designed to answer some very basic questions: How long does it take the arms to fall off? When does the skull start showing through? At what point is a body reduced to bare bone? It didn't take rocket scientists to realize that those processes occurred much faster in summer than in winter. Fairly quickly, though, our research projects became more sophisticated, and we developed timelines and mathematical formulas that could help us estimate, with surprising accuracy, how long someone had been dead once we obtained temperature records for the days or weeks prior to the body's discovery. The key, we learned, was "accumulated-degree-days," or ADDs—the sum of each day's average temperature. For example, if a body was placed at the Body Farm during the height of summer, when the temperature averaged a blistering 80 degrees each day, after ten days the body would have accumulated 800 degree-days—and would be well on its way to skeletonizing. During a ten-day stretch of brutal winter cold averaging a bone-chilling 30 degrees each day, a corpse would accumulate just 300 degree-days—and would have barely begun to bloat from the buildup of internal decomposition gases. The beauty of using ADDs to chart decomposition was that the data could be used anywhere in the world: by around 1250 to 1300 accumulated-degree-days, a body anywhere in the world would have been reduced to bare bone or bone covered with dry mummified tissue.

Then there were the bugs. One of our earliest research projects, conducted in 1981 by my graduate student Bill Rodriguez, charted the many insect species that came to feed on corpses: which bugs came, when they showed up, and how long they stayed. Perched for hours beside corpses, fending off flies that tried to lay eggs in his nose and mouth, Bill laid a cornerstone for what would soon emerge as the new specialty of forensic entomology. Today, thanks partly to Bill's pioneering insect study at the Body Farm, crime scene technicians all over the world know to collect insect specimens from the bodies of murder victims so entomologists can determine how long those bugs have been feeding on the flesh. Since Bill's groundbreaking insect study, numerous entomologists have come to the Body Farm, the only research facility in the world where on any given day, dozens of human corpses at every stage of decay—freshly dead, completely skeletonized, and everything in between—are there for the observing, as accessible to insects as they are to scientists.

Entomologists aren't the only scientists who rely on the unique research opportunities the Body Farm offers. One of my former graduate students, Dr. Arpad Vass, a research scientist at Oak Ridge National Laboratory, has spent the past several years sampling and chemically analyzing the gases given off by bodies as they decay. So far, Arpad has identified more than 450 different compounds in those gases; knowing exactly what those chemicals are is allowing Arpad to develop a mechanical "nose," an instrument he can program to sniff out clandestine graves, just as a cadaver dog can be trained to find bodies. Arpad has also used postmortem forensic chemistry, to coin an awkward term, to determine time since death: by analyzing the chemistry of death and decay products and studying how the ratios of various chemicals change as a corpse decays (just as entomologists study the changing parade of insect activity), Arpad can correlate chemistry with time, and read the decay products like a clock that has been ticking off the hours or days or weeks since death. He's also conducting research to understand why bodies give off energy fields; his hypothesis is that the chemical reactions of decomposition turn the body into a giant biochemical battery, in effect; if that proves to be the case, it means the Energizer bunny *could* retain some voltage even after thumping the bucket.

One thing most people don't realize about the forensic program at the University of Tennessee is that once a corpse has finished skeletonizing at the Body Farm, its scientific life, so to speak, has just begun. In fact, when people fill out the forms to donate their body to the Body Farm—as more than a thousand people have now signed up to do—what they're actually agreeing to donate is their skeleton; the flesh is just the biodegradable wrapper the skeleton arrives in. On a rapidly expanding set of shelves in locked rooms beneath Neyland Stadium, UT has built the largest collection of modern known skeletons (that is, of known identity, age, sex, stature, and race) in the United States, and perhaps in the world. By mid-2007, the collection—the William M. Bass Donated Collection, it's called—included nearly seven hundred specimens, with another skeleton being added about every three days, on the average. These specimens are a remarkable resource for training anthropologists and forensic scientists (besides our own students, the Anthropology Department helps train hundreds of crime scene and crime lab technicians every year, through the National Forensic Academy). They're also a gold mine of data for the Forensic Anthropology Data Bank, which amasses detailed skeletal measurements from people all over the world so forensic scientists confronted with an unknown skeleton can more easily determine the race and ethnicity of the bones: European, Native American, African-American, sub-Sahara African, Pacific Islander, Australian Aborigine, Chinese, or any of dozens of other groups included in the data. The donated skeletons also form the backbone of ForDisc, a powerful computer program developed at UT by my colleague and former student Dr. Richard Jantz to determine the sex, stature, and race of unknown skeletons on the basis of a few simple skeletal measurements. (ForDisc played a pivotal role in the case detailed in Chapter 9, "Listening to the Bugs.") During 2006 and 2007, every skeleton in the donated collection was scanned with a CT scanner. In the years to come, I expect those scans will be used for all sorts of interesting research and applied forensic science, such as the FBI's experimental facial-reconstruction software, ReFace (described in Chapter 14, "Leoma Patterson, Part 2").

One of the most dramatic and revolutionary advances in forensic science in recent decades is the advent of DNA testing. Although DNA testing isn't a magic wand—as the Leoma Patterson case makes painfully clear—it is an astonishing breakthrough. DNA research is no longer confined to the field of genetics; within anthropology, a new scientific discipline—"molecular anthropology"—is emerging. UT's anthropology faculty now includes a talented young molecular anthropologist, Dr. Graciela Cabana, who will doubtless find fascinating ways to advance the frontiers of her specialty through research at the Body Farm.

One piece of research that will probably never be done at the Body Farm is the effect of book writing on the health of the body, or at least the health of *my* body. In the year 2002, early in the writing of my memoir, *Death's Acre,* my heart stopped beating and I nearly died. Then, just as *this* book was nearing completion, my cardiologist informed me that my pacemaker—implanted after my 2002 brush with death—was dying, and needed to be replaced right away. I went in for day surgery on a Wednesday morning, and by lunchtime that day, I was headed home. The following day, I felt good enough to take my dog, Trey, for her afternoon walk, and early the next week I drove to Nashville and gave a two-hour lecture to a group of medical professionals. Officially I've been retired for years now, but some weeks I still put in forty or fifty hours of work—by choice, not of necessity. Occasionally I end up wishing I'd chosen to say no more often, but mostly I say yes because I love to lecture and love to consult on interesting forensic cases. Sometime soon, for example, I'm supposed to help a team of forensic scientists exhume and examine the remains of the famous magician Harry Houdini, who died on Halloween in 1926; he supposedly died from a ruptured appendix, but questions—and rumors about death threats and poison—have persisted for eighty years, veiling the truth like smoke and mirrors.

Houdini was arguably the world's greatest escape artist, yet in the end, he couldn't escape the Grim Reaper. None of us will, but some of us—thanks to the magic of technology and medicine—manage to prolong our performance by years. It's my good fortune that cardiac science, like forensic science, has made immense strides in the course of my adult life.

And yet: the human heart, like the human mind, remains mysterious and sometimes tragically flawed, as the unchanging penchant for murder reminds me again and again. It has been my calling and my privilege to help solve some of those murders, and—thanks to years of research by graduate students and faculty at the Body Farm—to provide scientific tools that help other forensic scientists solve them. I never set out to create something famous at Body Farm; I was simply putting one scientific foot in front of the other, trying to answer questions as they arose in the course of murder investigations or classroom discussions. Gradually, though, those research steps have taken me and my colleagues and students on quite a journey.

In the chapters that follow, you'll see how things we've learned at the Body Farm have helped us identify the dead, figure out what happened to them, and in many cases (though sadly not all) bring killers to justice. But the real breakthrough, as I'm reminded in the wake of the tragic shootings at Virginia Tech, remains elusive. The real breakthrough will come the day we learn not how to solve more murders, but how to prevent more murders.

Meanwhile, until that day dawns, our not-so-ivory-tower research, behind our locked gates and wooden fences, will equip investigators with more and better tools to solve the crimes that occur in the real world. The world beyond the Body Farm.

—Dr. Bill Bass

Knoxville, Tennessee

June 2007

1

THE GOLDEN BOWL, THE BURNING PALACE: APPLYING MODERN SCIENCE TO ANCIENT BONES

As fans of the television series *CSI* know, death scenes can capture a wealth of detail about what happens in the instant when human life is snuffed out—even, I can say with certainty, when that instant occurred nearly three thousand years ago.

More than four decades ago and six thousand miles away, I had one of my most memorable experiences in applying the tools of archaeology and anthropology to the questions of forensic science. The death scene lay in the ancient hilltop citadel of Hasanlu, in northwestern Iran, where a fierce army attacked the massive fortress, breached its mighty walls, and brought down its palace and temple in a rain of blood and fire. Hundreds had died in the battle and the blaze, but I was focusing on three of the dead, who were unearthed in a particularly dramatic discovery in the ruins.

Midway through the project, though, I began to fear that a fourth death might soon be involved: my own. As I lay doubled over, delirious for days on end, my circumstances may have been less heroic than those of the ancient warriors whose bones had drawn me here, but the setting—the way of life, the nearness of death, even the practice of medicine—had changed little in the twenty-eight centuries since the fortress fell.

In the summer of 1964, at age thirty-five, I was an eager assistant professor of anthropology at the University of Kansas at Lawrence. Although the ink on my Ph.D. from the University of Pennsylvania had been dry for only three years, by that time I'd excavated more skeletons than almost any other anthropologist in America. Starting in 1957, the summer after I entered graduate school at Penn, I'd worked for the Smithsonian Institution, which was excavating numerous Native American village sites throughout the Missouri River Basin. The Army Corps of Engineers was building a string of dams along the Missouri; the river's waters were rising; and the Smithsonian was racing to unearth and preserve as many relics and bones as possible before the sites were inundated forever. My first summer in the Missouri Basin, my crew and I painstakingly located and excavated several dozen

Arikara Indian graves; by 1963, after I'd devised a way to use road-grading equipment to peel back the earth atop the graves without damaging the bones within, we were excavating several hundred each summer. I got so fast and efficient at excavating burial grounds that I eventually earned the label "Indian grave robber number one" from a Native American activist who disapproved of digging up Indian graves. (This chapter in my career is told in more detail in my memoir, *Death's Acre*, published in 2003.)

So although I was still relatively young in 1964, I'd racked up some unique and extensive experience, and it didn't come as a total surprise when I got a call from Bob Dyson, an up-and-coming archaeologist back at Penn, seeking help excavating ancient graves at Hasanlu. Millions of people are familiar with the work of Egyptologist Howard Carter, the British archaeologist who discovered the tomb of King Tut. Far fewer are acquainted with Bob Dyson's work, and that's a shame, because Bob was the Howard Carter of Hasanlu. Six years before inviting me to Iran, when he was a mere thirty-one years old, Bob made the discovery of a lifetime at Hasanlu. Beneath a mound of rubble and charred embers in the wreckage of an immense two-story palace or temple, he found the skeletal remains of three men trapped—crushed—when the burning roof collapsed and the massive walls toppled. As the debris was carefully brushed from the buried bones, it became clear that the three men had died while at a dead run, literally, their arms and legs frozen in perpetual near-motion. It was a remarkable snapshot of death, preserved for nearly three millennia.

Even more remarkable—and a key reason Bob invited me to Hasanlu—was the object cradled in the arms of the front runner. The object was a bowl (or a vase, or a beaker): a metal vessel measuring about eight inches high, seven inches across the top, and six inches across the base. The falling walls had flattened the bowl, of course, along with the guy carrying it. Even so, the bowl's elaborate ornamentation remained virtually intact and astonishingly detailed. An upper scene, enacted by a band of embossed figures encircling the bowl, showed three young men bringing offerings to the gods—two gods riding in chariots, and a third god wearing a horned headdress. The bowl's lower ring contained a series of smaller scenes and numerous figures, including a nude goddess, a male hunter or warrior, an eagle carrying a woman, another woman riding a lion, and a trio of people—a seated man, a woman, and a child, whom the woman is presenting to the man. The bowl was such an extraordinary find that *Life* magazine devoted an eleven-page spread to it—the equivalent, back then, of a one-hour television special hosted by Katie Couric or Diane Sawyer.

Despite its relatively small size, the bowl was quite heavy. That's because it was made of solid gold. Today—carefully guarded in a museum in Tehran—the bowl is one of Iran's richest archaeological treasures. Three millennia ago it was equally prized, occupying a sacred place in the citadel's palace or temple—as valuable as a freshly carved or painted masterpiece by Michelangelo would have been to a Renaissance pope or a Medici prince. Three men, at least, had died for the bowl's sake—caught for all time in a dramatic act of desperation.

By the time Bob Dyson called me in early 1964, millions of Life readers had marveled at the bowl's intricate beauty. But the question that still hung in the air, across all the centuries since the citadel's gates were breached, the roof caught fire, and the walls collapsed, was this: Who were these three men, and what were they doing with the bowl? Were they all citadel defenders, frantically trying to keep a sacred relic from falling into the hands of infidels? Or were they looters, greedily racing from a burning temple with the richest plunder of all? Or were they a mixture of both: one lone temple guard, faithful to the last, fleeing a pair of ruthless pursuers?

Bob asked if I could help answer the question. I thought I could, but it wouldn't be easy. I'd need to travel to Iran, dig up the bones of soldiers from both ancient armies, and compare their measurements to those of the area's modern inhabitants. It was a long shot, back in those decades before scientists knew how to use DNA to trace ancestry, and it had never been done before. And it was irresistibly interesting.

Bob promised to provide me with a crew of ten local workmen to do the digging, and offered to pay travel expenses for one of my students, to help supervise the workers. I invited Ted Rathbun, an inexperienced but promising student who would be starting graduate school in the fall. I thought I was doing Ted a favor. Little did I know that soon my life would be in that young man's hands.

Ted and I left Kansas City in early June aboard an eastbound TWA 707; we made stopovers in Washington, D.C., to visit my mother, and New York City, to take in the World's Fair. (In those days, airlines didn't charge extra for making stopovers between legs of a long flight.) From New York we flew to London, where we stayed long enough to make a side trip to Stonehenge, which had already been standing for a thousand years or more when the citadel at Hasanlu fell. While in England, we also visited with two of the scientists who'd exposed the infamous "Piltdown Man" hoax, in which some clever prankster created an evolutionary "missing link" by burying together pieces of the mandible (lower jaw) of an orangutan, teeth from a chimpanzee, and the jawless skull of a medieval human.

Having thus broadened Ted's horizons, both geographically and archaeologically, we set off for the Middle East, arriving in Beirut, Lebanon. In those days Beirut was a beautiful, vibrant, cosmopolitan city, brimming with visitors from many nations. You could ski in the nearby mountains in the morning, then unwind with an afternoon swim in the Mediterranean, gazing above the blue waters at a sky filled with parasailors. There was no gunfire, no explosions; this was a decade before a long, bloody civil war tore Lebanon apart. Beirut was a last, urbane vestige of modern civilization before the wilds of Iran.

We'd flown into Beirut on TWA; in Beirut we switched to Cedar, a Lebanese airline, for the flight to Tehran. There, we rendezvoused with Bob Dyson and the rest of the American team who would be spending the summer in Hasanlu. From the airport, we were whisked straight to a dinner at the American embassy, where we were treated like visiting dignitaries rather than lowly academics and students.

Compared with Beirut, Tehran struck me as far more conservative and militarized. Soldiers patrolled everywhere, checking passports and other belongings. Iran was still ruled by the pro-American shah, so Tehran seemed relatively modern and Western, but here and there I saw signs of the Islamist fundamentalism that would eventually transform Iran into a militant Muslim state.

We stayed in the Iranian capital for nearly a week, getting our visas, passes, and other paperwork processed. In planning the expedition, Bob Dyson figured he might want to send some of the crew to other nearby sites, so he had requested driver's licenses for me and several others. The request was refused—the government official who vetoed it didn't say why. However, Bob had worked enough summers in Iran to know how the machinery of corruption and bribery worked. He handed over some dollars to lubricate the wheels of progress, and soon we had our licenses.

Official corruption wasn't the only problem I encountered in Tehran. My first inkling of illness came there, when I developed a case of diarrhea. It wasn't bad—it happens to travelers all the time—and I took some Imodium tablets, which seemed to help.

After our work in Tehran we boarded a smaller propeller-driven plane and flew several hundred miles northwest to Tabriz, the provincial capital of East Azerbaijan. Our first stop was the U.S. consulate (a branch office of the embassy). The American consul in Tabriz was a young State Department staffer named Carlton Coon Jr. His father, Carlton Sr., an anthropologist who specialized in the races of modern man and of fossil man, had served on my doctoral committee at the University of Pennsylvania between 1956 and 1960. Not surprisingly, the young diplomat took a keen interest in this expedition from his dad's university. As it later turned out, that interest may have saved my life.

After a couple of nights in Tabriz—more than two weeks after leaving Kansas—we finally hit the road for Hasanlu. Hasanlu was a hard day's drive from Tabriz, so we took full advantage of being there to stock up on supplies. Bob Dyson's equipment included a one-ton flatbed Ford truck with wood-stake sides—the sort of truck that could be seen all over Kansas hauling hay bales and livestock across the prairie. Bob crammed the back of the truck with food, tools, and Ted and me, and off we went in a cloud of dust. The roads weren't paved but they were gravel, and they were pretty good. They'd been built by American troops during World War II to transport lend-lease equipment to Russia—convoys of trucks and tanks and other heavy supplies—and twenty years later, they were still good. Dusty, but good.

Much of the drive took us along the shores of Lake Urmia, which resembles Utah's Great Salt Lake in both size and mineral content. Both are inland lakes that have no outlet, so as water evaporates, mineral deposits that have been carried down from the region's mountains get left behind. As we threaded the shores of Lake Urmia in a cloud of dust, we saw local people wallowing in the mineral-rich mud flats, which ease the symptoms of rheumatism and other ailments.

One of the rivers that feeds Lake Urmia is the Solduz. By the time it reaches the lake, the Solduz has been tapped nearly dry by irrigation. Rainfall is rare in the summer; most of the annual precipitation comes in the form of winter snows, which accumulate on the higher surrounding mountains—some of them reaching 10,000 feet or more—and feed the rivers as they melt during the hot summers. The region's hillsides are a stark grayish-brown, but the irrigated valleys are lush and fertile, thick with wheat, fruit, nuts, and rice. We weren't in Kansas anymore, but if I ignored the mountains in the distance and focused on the rippling fields of wheat, I could almost forget that, for a moment at least. Then some unfamiliar sight would remind me I was thousands of miles (and many centuries) away from home: small children herding immense water buffalo; a towering haystack ambling along the road, completely enveloping the donkey that carried it.

The modern-day village of Hasanlu—and I use the term "modern" very loosely—was home to five or six thousand people, most of whom worked in the agricultural fields lining the Solduz Valley. Situated at an elevation of around 4,500 feet, the village would heat up into the nineties during the day but cool off considerably at night, as high desert locations do. By the time we bumped to a halt that evening, the sun was sinking, the heat was relenting, and Ted and I were caked with enough dust to render us nearly as dark-skinned as the villagers.

As best I could tell, Hasanlu had changed little since 800 B.C. The chief mode of transportation was walking, with donkey-riding a distant second. Buildings were made of mud bricks, baked in the sun; their roofs were constructed of saplings and branches laid across the top of the walls, then covered with several inches of packed dirt. In winter, it was important to sweep the snow off the roof, lest the weight cause the roof to collapse, or the spring thaw turn the dirt to mud. Inside the houses, the floors were dirt as well. The one exception to these ancient building codes was the village schoolhouse, which boasted a floor of concrete and walls of kiln-fired brick. The houses also tended to have haystacks on their roofs; if one catches fire, it's easy for a whole village to go up in flames.

The town was strung out along the irrigation canal, called a "jube" (rhymes with "tube"). The jube was more than just a source of irrigation water; it was also where livestock came to drink, where women came to do laundry and wash dishes, and where people gathered to gossip or to pick up prostitutes, who were nicknamed "jube queens" for their habit of sitting alongside the jube and dangling their feet in the water while waiting for customers. The whole scene was like something from the days of Jesus or Mohammed, but then my eye caught a flash of familiar modern images: beneath a cloud of flies, the tables in the town's open-air meat market were topped with flattened beer cans—Budweiser and Black Label empties. The surprise and incongruity of it made me smile.

One other note of modernity was a generator, which the American archaeologists had brought, and which ran every evening from six to ten. That one piece of technology changed the village culture dramatically. In the centuries B.G. (before generator), people went to bed at dark; in the A.G. era, people sat up at night for hours in front of the artificial breeze stirred by small electric fans, the village's new status symbol.

I was glad we were spending the summer, rather than the winter, in Hasanlu. Winters are harsh, and the main source of heat is animal dung. It's the children's job to gather dung and mix it with straw into bricks or cakes (a word that seems terribly wrong paired with "dung"!). After they've been dried in the sun, the dung cakes are stacked into pyramids or cones, some of them the size of houses. In winter, these structures would be covered with snow; to get fuel, villagers would tunnel deeper and deeper into them, removing the cakes from the inside out. It would be important, I suppose, not to remove too many bricks from a key structural spot lest you cause the massive layer cake of dung to collapse on yourself. Even in the summer, dung cakes remain a primary fuel source; to heat a couple dozen buckets of water for my crew to shower, for instance—one bucket of soapy water and one of rinse water per person—required 2.5 dung cakes, at twenty-five cents per cake.

Because it was summer, the village school was vacant, so it had been converted into a dormitory for the visiting archaeologists and anthropologists, who numbered only ten or twelve altogether. Three of the group were women, and for the sake of propriety, they stayed in a separate wing of the school, separated from the men by a wall with a locked door. Ted and I bunked together in a classroom, which was quite small—maybe eight by ten feet. The room had only one tiny window, and our only source of light was a Coleman lantern, so it was very dark inside; early on, we whitewashed the walls to offset the gloom.

The archaeological site lay just outside the village. The citadel was roughly circular in shape, and about the size of the University of Tennessee's Neyland Stadium, one of the largest stadiums in America. The fortress had been built atop a mound rising above the valley floor, to make it harder to attack. During World War II, Allied forces put artillery emplacements on the mound; I don't know who they thought might attack here—maybe Rommel's tank corps, which was sweeping across Africa and toward the Middle East. Twenty-eight centuries before Rommel's panzers, though, the only armored vehicles were chariots, and in ancient Hasanlu, a chariot road angled up the slope from the valley to the main gate. The wall around the perimeter was about twenty feet high; defenders could be arrayed along the top of the wall, with archers also firing arrows from four large square towers spaced around its wall. Inside the walls, the palace complex was huge, with walls that had soared some seventy feet high. In previous summers, Bob's crews had partially reconstructed the walls to stabilize them.

To begin to answer Bob's question about who the three men with the bowl were, I needed to see and measure as many skeletons of citadel defenders as possible.

Bob had hired about a hundred local men as workers for the summer; he would use ninety of them to dig for artifacts; Ted and I were given thirteen to excavate buried warriors. My assumption—an educated guess—was that the citadel's defenders would be buried in the local cemetery, while invaders killed in the assault would not be. In the course of my Native American digs in South Dakota, I'd grown accustomed to using earthmoving equipment to remove tons of topsoil quickly. Here, my resources were more limited; instead of road graders, I had the Iranian prototype of a dump truck: a donkey with a couple of jute sacks we could fill with dirt or rocks. The division of manpower in my crew was simple: four pickmen, four shovel-men, four wheelbarrow-men, and one water boy. Still, despite the primitive approach, we developed a good rhythm—the pickmen were surprisingly skilled at reading the soil and avoiding damage to the ancient skulls and bones—and soon we were unearthing several skeletons a day. Over the weeks to come, we would excavate a total of eighty-three burials, which was a dozen more than crews had excavated in the previous six summers combined.

As the skeletons mounted, we began to get a good idea of the skeletal characteristics of these ancient warriors. Their bones tended to be quite robust, with prominent muscle markings created by the tug of mighty sword arms and powerful legs. That makes sense: fighting men need to be big and strong. The three men associated with the golden bowl fit this general description as well, with one notable exception: the man who died with the bowl in his arms was big all right, but judging by the very slight muscle markings, he couldn't have been very strong. Perhaps his size got him a job as a palace guard, but, having a more sedentary job than a soldier would, he got soft. Ted went so far as to speculate that he might have been the palace eunuch, an unprovable but reasonable hypothesis, since castration would have reduced his testosterone level drastically, and testosterone helps athletes build muscle mass (one reason some hard-core female bodybuilders take synthetic testosterone). I was pretty confident that, given his lack of strength, the guy carrying the bowl wasn't a battle-hardened invader who'd fought his way through the ranks of the citadel's defenders.

But what about the two other guys hot on his heels—were they running *with* him, watching his back? Or were they running *after* him—chasing him, and on the verge of catching him when the walls came tumbling down? To answer that part of the question, I needed to measure Hasanlu's living, breathing inhabitants and see what clues to the dead I might find among the living.

Before leaving Kansas, I'd searched the literature and found that there were no modern studies of the people of the Solduz Valley. An anthropologist named Henry Fields had measured several populations in the Middle East, including groups about seventy or eighty miles south, but none in Azerbaijan, and certainly none in the Solduz Valley. I asked Bob Dyson if I could measure workers during lunch breaks or other slow times, and he said sure. So I'd go around and measure people who were on the crews—their stature, arm length, cranial length and breadth, and the height and width of their noses (which tended to be big and beaklike). After a few days I noticed that people were lining up to get measured—not just workmen but women and children, who were walking out from the village, a thirty-minute round-trip. This puzzled me greatly, so I asked the interpreter to find out why my measurement project had made me so popular. It took some asking around for him to find out the truth, but eventually he learned the reason: word on the street had it that I was measuring people for overcoats for the winter, and that everyone I measured would get a free coat.

By that time I'd measured sixty or eighty people, and I felt bad that, through no fault of my own, people's hopes were going to be raised and then—come winter—dashed. I tried to correct the misinformation;

I asked the interpreter to explain that this was strictly science, that nobody from this area had ever been measured before, and that I wasn't going to be able to send overcoats. But the villagers simply wouldn't believe that, and they kept coming, so I stopped taking my measurements.

Another reason I stopped taking measurements was that I got deathly ill. The intestinal bug I'd picked up in Tehran came roaring back with a vengeance, causing me to double over with cramps. My digestive system didn't know what hit it—bouts of near-constant diarrhea alternated with spells of painful constipation. I was weak from hunger and dehydration, because I was vomiting frequently, too. One moment I would be shaking with chills, and the next, sweating with fever. I spent days in delirium.

During the day, while Ted was out supervising the excavation crew, the expedition's recorder, Carolyn Dosker, kept an eye on me, bringing me boiled water, yogurt (the Iranian cure-all), and a bit of rice when I could handle it. At night, Ted—after a long day of running herd over the crew—would take over. During my lucid moments, I thought of my wife Ann, who was eight months pregnant; my sons Charlie and Billy; and the baby I might never see. I vowed not to die in Iran.

I got help keeping that vow. I'd written home when my illness began to worsen, asking if my physician could send me some prescription antibiotics, since the over-the-counter medications we had weren't helping. Unfortunately, it wasn't possible to send prescriptions through normal channels. Luckily, thanks to Carlton Coon, we had access to other, better channels. Just when I was starting to wonder whether I would survive, a diplomatic courier pouch reached Tabriz from the States loaded with powerful antibiotics, which were brought to me in Hasanlu. The medicine didn't cure me overnight, but in the days that followed I rallied, regained much of my strength, and returned to the field. It would take me another six months to recover fully, but I knew within days of the medicine's arrival that I wouldn't die in a primitive village in the Solduz Valley, thousands of miles from my home and my family.

After I was back on my feet, we took our crew to spend several days excavating at a Kurdish site—a mountaintop fort that had been destroyed by an earthquake a couple thousand years ago. The drive over was hot and dusty, so we stopped at a creek to cool down and clean up. We stripped to our boxer shorts to bathe in the creek; I looked up from the water and noticed a crowd of Kurdish women gathered on a nearby hill to look at us. We waved and they chattered excitedly, but I'm not sure what they were saying.

The ruins of the Kurdish fort were terraced into a mountainside about a thousand feet above a valley. During our lunch break, we wrestled a big rock to the edge and rolled it off, then watched and whooped as it crashed down a thousand feet. From the pile of debris it shattered upon, I could tell we weren't the first to succumb to that boyish impulse.

Back at Hasanlu, we began to wind down for the summer. We'd been boxing up the bones all along, as they were excavated. Ted and I left before the expedition completely wound down for the summer. We left maybe a week or two before Bob did—school was starting soon, and we had families to get back to. Ann had given birth to our third son, Jim, and she and I were both eager to have me back home. Bob had to shut down the site and sort through the summer's worth of excavated relics. We just picked a day and stopped.

At the end of the season, when the site was shut down and the crews were all gone, the U.S. ambassador and the head of the Tehran museum would inspect the relics, which had been divided into two batches of roughly equivalent value, labeled "A" and "B." Small slips of paper would be put into a hat, and the American and the Iranian would draw lots to see which nation got batch "A" and which got "B." We'd divided all the skeletal material into sixteen roughly equal units—big wooden crates of skeletal material—sitting out in the hallway of the museum. But either no one cared about the bones or they never got around to dividing them up, because those bones eventually trailed me back to Kansas, and later from Kansas to Tennessee.

But I get ahead of myself.

In mid August of 1964, I got to put my Iranian driver's license to good use. I drove a Land Rover carrying Ted and a couple of other anthropologists from Hasanlu to Tehran. From there I flew to Beirut, then London, New York, and Kansas City, where I found Ann, Charlie and Billy, and baby Jim waiting for me.

I'd set out strong and confident; I arrived home weaker and less certain. Besides falling gravely ill, I'd encountered two serious problems in answering Bob Dyson's question about the three men with the bowl. We knew that the attackers had burned down the palace, but we didn't know if they had taken possession of the village and stayed put, or had simply destroyed the citadel and moved on. That was one problem. The other was a recent and drastic change in the region's migration patterns. For thousands of years, the population in the Solduz Valley had remained remarkably stable and isolated. Unlike Europe, where populations have long mixed and mingled, this area was remote and rugged enough to remain geographically and genetically insulated. Lake Urmia provided a water barrier to the northeast; the Zagros Mountains reared thousands of feet high to the west, limiting contact with Iraq; and even higher mountains to the north (including 18,000-foot Mount Ararat) blocked migration from Turkey. But in the 1900s, roads and powered vehicles breached these barriers as surely as Hasanlu's ancient attackers had breached the citadel's walls and chariot gate. Many of the modern inhabitants of the Solduz Valley had come down from Turkey during the past century; when I took the measure of some "foreign"-looking head or nose, I didn't know if I was recording a descendant of some ancient invader or the grandson of a twentieth-century transplant. Today, as anthropologists learn to use DNA to trace the ancient roots of modern populations, it might be possible to provide a more definitive answer to the identities of the three men with the bowl. Possible, but expensive.

And I believe the answer would be the same as what my intuition told me back in 1964 as I pondered the death scene in a village where time—like the dead men—had been frozen for three millennia. I believe the soft-bodied guard with the bowl cradled in his right arm was in the company of two stronger, tougher palace guards: fighting men handpicked for their size and strength. The one least able to fight was given the bowl to carry; the other two were watching his back as they raced to save the bowl from fire and invaders.

I've seen scores of death scenes since then. Each one has told me a story, sometimes complete, sometimes fragmentary. Many have been more gruesome; none has been more memorable—or more tantalizing—than the 2,800-year-old tale whispered down through the ages by the golden bowl in the burning palace.

2

SPLASH LANDING, PART 1: FINDING ANSWERS IN TEETH AND SKULL MORPHOLOGY

One of the main missions of a forensic anthropologist is to restore names and identities to the unknown dead. Often these are murder victims; sometimes—as in the aftermath of Hurricane Katrina or the World Trade Center attacks—they are victims of mass disasters (mass murder, obviously, in the case of 9/11); occasionally they are people who died of natural or accidental causes and were found without any form of identification.

"Identity" is an interesting notion. What is it that makes us unique as individuals? Is it the sum total of our thoughts and feelings? Is it our place in a family tree, as offspring of a particular set of parents? Or the paper trail we create as we buy cars and houses, cash paychecks, file tax returns? The patterns of raised loops and whorls that help our fingertips grip objects more securely? The chemical composition of our DNA?

The answer, of course, is "all of the above," and more. Forensically speaking, it's possible to make a positive identification of an individual in a variety of ways. In recent years, the gold standard has become DNA typing: each of us carries, within the nucleus of every cell in our body, a genetic blueprint written in biochemical code—a coded message that's three billion lines long, and different from every other person's. But DNA testing is far from the only technique for positive identification, and it's often not the fastest or most efficient. Despite what you might gather from television shows like *CSI,* DNA analysis can take weeks or even months and can cost thousands of dollars. Genetic testing is getting faster and cheaper, but it still has a long, long way to go to rival the speed and affordability of matching a skeleton's teeth, fillings, and other unique features with missing persons' antemortem (before-death) dental records, X-rays, and other medical information.

One afternoon in March of 1974, not quite three years after I moved to Knoxville to head the Anthropology Department at the University of Tennessee, four people showed up in my office beneath Neyland Stadium: the sheriff of Blount County (the county just south of Knox County), the coroner of Blount County, the president of the East Tennessee Pilots Association, and a reporter from the *Maryville Daily Times.*

Actually, four people and one skull showed up in my office. The skull had been found, partially buried in the sand, by a boy walking along an exposed stretch of shoreline in Blount County. That explained the newspaper reporter: Maryville is the county seat of Blount County, but it's a small town, so it's big news when somebody finds a human skull along a freshly exposed stretch of riverbank.

Every winter the Tennessee Valley Authority (TVA) lowers the water level in the string of lakes it has created along the Tennessee River and its tributaries, as a flood-control measure so that when the spring rains set in, the reservoirs can accommodate the extra runoff. Some of the lakes on the Tennessee's tributaries—Norris and Fontana lakes, for instance—are lowered by fifteen or twenty feet every winter; Fort Loudoun Lake, though, must remain navigable by barge traffic, so TVA drops it by only a few feet. Even so, a drop of several feet is enough to expose large stretches of shoreline—prime territory for arrowhead hunters looking for souvenirs of the valley's prior, pre-TVA inhabitants.

I don't know whether this Blount County boy was looking for Indian relics; in any case, what he found was a human skull. He left it there by the river, but that evening he told his father about it. "We'd better go back and get it," said the boy's dad. Once they'd retrieved the skull, they contacted the Blount County Sheriff's Office, which was understandably anxious to determine whether the skull was ancient or modern, and whether it belonged to a murder victim.

From the heavy, prominent brow ridge over the eyes, this was clearly a male skull. The narrow nasal opening and the vertical structure of the teeth and jaws told me that it was Caucasoid rather than Negroid (in black or African-American people, the teeth and jaws tend to angle forward, a distinctive trait called "prognathism"). And the fact that the incisors were basically flat across the back, rather than shovel-shaped, ruled out the possibility that this was a Native American skull. Besides, except for the lack of a mandible, or lower jaw, the skull was in remarkably good shape—far too pristine to have been stuck in the mud for decades or centuries. Amalgam fillings—done by a good dentist, judging from the handiwork—confirmed the skull's recent vintage. The skull was unblemished; if it belonged to a murder victim, the murder was definitely not committed by a blow or gunshot to the head. From the wear on the teeth and the prominence of the cranial sutures—the joints between the bones of the cranial vault—I estimated the age to be somewhere between thirty and thirty-four.

Only two things about the skull appeared to be at all out of the ordinary. One was the presence of adipocere, a waxy or soapy substance, at the back of the eye orbits. Adipocere—the Latin term translates literally as "grave wax"—forms when fatty tissue decays in a very moist environment, such as a basement . . . or a riverbank. Most people don't realize it, but all of us except for the extremely malnourished have pockets of fat behind our eyeballs. If you've ever seen photos of concentration camp survivors, you'll notice that their eyes appear sunken, extremely deep-set in their skulls; that's because they've used up all of their fat reserves, including those small pads behind the eyes. So knowing where the skull had been found, I wasn't surprised to see the clumps of adipocere deep in the eye orbits.

The skull's other unusual feature was its shape: the cranial vault was remarkably tall and narrow, with a sloping forehead and slanted openings, or orbits, for the eyes, which gave the skull an almost Oriental appearance. What immediately sprang to my mind when I saw the high, narrow vault was a rare medical condition, a deformity of the skull called "scafoscephalae," which results from premature closure of the sagittal suture, the joint between the frontal bone (the forehead) and the parietal bones (the sides of the skull). That means that during childhood the skull can't grow wider, so it grows higher instead, to accommodate the brain's growth. If you take the top of the skull from someone who had scafoscephalae and turn it upside down, the inside of the cranial vault resembles the V-shaped inner hull or keel of a boat. Today, when children are born with this disorder, it's correctable: using a Stryker saw, a surgeon can reopen the sagittal suture, allowing the skull to grow normally; eventually the cut bone will fuse together again, but not before the skull has grown to a normal width.

Now, this particular skull wasn't quite tall and narrow enough to merit a diagnosis of scafoscephalae—if this individual were a child today, he probably wouldn't require surgical intervention—but its proportions were much closer to scafoscephalic than most people's.

Of the four people who accompanied the skull, the most unexpected was the president of the East Tennessee Pilots Association, Jim Cline. Cline told me about a Knoxville man named Elmer Reynolds who had disappeared two years before in a small plane—Cline's plane, in fact—while taking a sixteen-year-old girl for what was supposed to be a brief sightseeing flight. The girl, Linda Hendrick, was from Michigan; she was visiting an aunt in Knoxville—an aunt who lived with her husband in the caretaker's trailer at Cox's Skyranch, a small grass airstrip tucked alongside the river about four miles upriver from where the skull was found. When the plane took off near dusk one afternoon in January of 1972, Reynolds said he was going to show Linda the lights of Knoxville. It would need to be a short flight, as the plane had only about a half hour's worth of fuel on board. When they didn't return, the Civil Air Patrol launched a search from the air but failed to spot any sign of aircraft wreckage in the wooded terrain or the river shallows; the local rescue squad did some dragging in the river, also without finding anything. One theory that circulated among some law enforcement officers was that perhaps Reynolds, age thirty-two—a man with a wife and five children—might have run off with the teenager. But inquiries at nearby airports where he could have refueled the plane also failed to turn up any signs of its whereabouts or flight path.

The skull was the right age and race to be that of Reynolds. It was obvious from the good condition of the skull's teeth that this man had received regular dental care. What's more, one of the molars had a distinctive shape and a distinctive filling.

Most people's molars have what's called a "Y-5" pattern; that means the chewing surface of the molar has five peaks, or cusps; the name comes from the fact that the valleys between the five cusps connect to form the shape of the letter Y. One of the skull's molars had a less common pattern, a "4-+," in which the valleys between four cusps connect in the shape of a plus sign. These different patterns in molars date back millions of years in our evolutionary history—all the way back to our ancestors, apes of the Miocene epoch (about 27 million years ago). Between the Miocene epoch and the twentieth century, dentistry evolved even more rapidly than primates did. By the 1970s, highly evolved dentists knew that if a patient developed a cavity in the central valley of a molar, it wasn't enough just to fill the hole itself; a lasting repair required drilling out along the valleys a ways and filling a larger area to make sure the filling was anchored to a solid foundation. This guy, who had a long, uniquely shaped filling in a 4-+ molar, clearly had a highly evolved dentist. "I'm sure Elmer Reynolds' dentist could tell us if this is him," I told the small group in my office. One of them—I forget who—called Elmer's wife, who steered us to a dentist named Robert Greer.

I called Dr. Greer, who had an office in an old house on Magnolia Avenue. "I may have the skull of one of your patients," I told him. "Could I bring it out and get you to take a look and compare it to your records?"

He agreed, and asked me to bring the skull to his office early the next morning.

I showed up at 9 A.M. with the skull in a brown paper grocery bag. "This might be Elmer Reynolds," I said. Dr. Greer reached into the bag, and as he lifted out the skull, he said at once, "Yeah, that's Elmer." He said it before consulting Reynolds' dental records; in fact, he said it even before inspecting the teeth. He, like me, was struck immediately by the skull's high, narrow vault—a vault the dentist had bent over at six-month intervals for four years. If Greer had wondered why Elmer hadn't come in for his regular cleanings, his question was now answered.

Dr. Greer, it turned out, had served as a dentist in the Air Force before setting up private practice as a civilian. During his years in the military, he'd been called on several times to make dental identifications of pilots killed and burned in plane crashes, so he had both interest and experience in forensic dentistry, or odontology. Dental identification—matching a dead person's teeth to antemortem (before-death) X-rays or records—has been used as evidence in court for more than a century, but the armed forces are the nation's main repository of odontologists, because soldiers are at greater risk of dying in explosions, plane crashes, and other settings that can make ordinary visual identification impossible.

After recognizing the distinctive shape of the skull, Dr. Greer pulled out a chart and confirmed that the teeth matched Elmer's dental records. The dentist and I swapped stories about identifying plane crash victims, then I thanked him for his help, put the skull back in the paper bag, and walked out of his office.

As I headed down the front steps toward my car, I passed two men on the sidewalk. We said hello in passing, and then one of them said, "Excuse me, are you the man trying to identify our brother?" Evidently they'd heard a skull had been brought to me for identification, and when they called my office my secretary told them I'd gone to consult Dr. Greer.

I told the men who I was, and acknowledged that I was trying to identify a skull found on the edge of the reservoir. Then they introduced themselves, and their last name was Reynolds, so I figured they probably were indeed Elmer's brothers, and deserved to know where things stood. "I hate to tell you this," I said, "but we have made a positive identification, and it is Elmer Reynolds." We talked for a few minutes, and then I noticed the men glancing at the bag in my hand. Now, as far as I know, neither Emily Post nor Miss Manners ever offered any etiquette advice about this particular sort of situation, so at this point I was forced to wing it. After a moment's thought, I said, "Would you like to see your brother?" They discussed it for a few minutes and finally decided they would. I took the skull out—this was on a sidewalk along Magnolia Avenue, a busy Knoxville street—and they both said, "Yeah, that's Elmer all right." I found it interesting that people who knew him could recognize him even without the soft tissue.

I never had any contact with Reynolds' widow, but I heard she was relieved that we had made an identification, which meant she could finally file a death claim and collect on his life insurance. With five children to support, she must have had quite a financial struggle over the past two years. And although there's nearly always sadness when a body or skeleton (or skull) is identified, there's also closure, and the opportunity—painful though it may be—to move from uncertainty and confusion to grief, resolution, and the process of getting on with life again.

I'd received the skull on a Monday. On Friday, I sent the Blount County authorities my report detailing my examination of the skull and the positive identification, based on a distinctive filling in one of the skull's molars, which precisely matched Elmer Reynolds' dental records (as did an unusually shaped root in one of the teeth).

The next week, I clambered into a flat-bottomed aluminum boat belonging to the Blount County Rescue

Squad. They beached the boat in a small, sandy, debris-laden inlet tucked between rocky bluffs lining that stretch of the river: the site where the boy had found the skull. While rescue squad divers probed the depths and volunteers probed shallower waters with aluminum poles in search of the telltale clatter of metal wreckage, I sifted through sand and layers of washed-up twigs and trash hoping to find bones from Elmer's postcranial skeleton (the portion below the skull) and anything from his passenger, Linda Hendrick.

The rescue squad's efforts expanded; six other area rescue squads joined the search, extending it for miles upriver. But I found nothing more where the skull had washed up. In hundreds of hours of searching, the divers and pole-probers found a tool bag Jim Cline said had come from the plane. But that was it: no aircraft, and no more bones. The river was not willing to give up its dead. Not yet, at any rate.

And that, it seemed, was that.

Until thirty-one years later, when the search resumed.

But that's a story for another, higher-tech chapter in the pages that follow.

3

SHEDDING LIGHT ON A VICTIM'S BONES: USING UV TO ILLUMINATE IDENTITY

One of the most revolutionary changes in forensic science in recent decades has been the advent of DNA testing: the ability to chart any person's genetic makeup—to take a genetic "fingerprint," essentially—and compare that with all sorts of forensic evidence, ranging from body fluids (including blood, saliva, and semen), to hair, to skeletal elements such as soft tissue, teeth, and bones.

DNA analysis isn't without its drawbacks. For one thing, it can be time-consuming and expensive, requiring weeks or even months to process some samples and costing as much as several thousand dollars for a single comparison. For another thing, it's not always possible to obtain DNA samples, even if you have a complete skeleton; immersion in water, burial in bacteria-rich or acidic soil, or incineration at temperatures hot enough to reduce a body to brittle, calcined bones can make it impossible to recover enough genetic material for a DNA analysis. Still, it's a remarkable and powerful new tool in the forensic scientist's toolbox, and I wish it had been available throughout my career.

More specifically, I wish it had been available back in 1975, when I received a box of human bones that were found in a barren field in Kansas. But it wasn't, so instead I used a combination of ultraviolet radiation and chemistry to shed light on the case of a young murder victim. Thirty years later, I took the witness stand and testified about the technique in the trial of a man finally charged with murder in that case.

The box, which arrived by certified mail on January 20, 1975, came from Dr. James Bridgens, a Johnson County medical examiner in Shawnee Mission, a suburb of Kansas City. I knew Dr. Bridgens slightly from a case I'd worked several years before, in which I managed to identify Lisa Silvers, a two-year-old murder victim, after finding her distinctively notched front teeth submerged in a sandy creek bed outside Olathe.

The bones inside the box from Shawnee Mission had been stumbled upon two weeks earlier, scattered in a field outside Lenexa, about seven miles away. Today,

Lenexa is all subdivisions and shopping centers, but in 1975 it was a sleepy farm community whose biggest boast was that it had been "the "Spinach Capital of the World" in the 1930s.

I unpacked the box and inventoried its contents with the help of one of my Ph.D. students, Doug Owsley (who is now the senior physical anthropologist at the Smithsonian Institution). The box held just fourteen bones, or less than 10 percent of a human skeleton: six ribs, two vertebrae, three long bones (a left femur and tibia and a right humerus), a right ilium (part of the hip), a right scapula, and—fortunately—the skull, minus the mandible.

Dr. Bridgens had already matched the teeth and fillings that remained in the upper jaw with the dental records of Lizabeth Wilson, a thirteen-year-old girl from Shawnee Mission. Lizabeth had gone missing the previous July; she was last seen walking home from a public swimming pool near Shawnee Mission East High School. According to the story I heard, Liz's older brother was supposed to walk her home, but he went off with some friends instead, leaving her to make the walk alone. She vanished without a trace.

Since Dr. Bridgens had already matched the teeth and fillings in the skull with Liz's dental records, he didn't need my help in identifying the girl. He needed help making sure that Liz was the only victim whose bones we had. On the surface, it might seem obvious that they were all hers, but when you've worked as many murder cases as I have, you learn that what's *obvious* is not always the same as what's *true*. Serial killers sometimes dump multiple bodies in the same location; in Knoxville's most notorious serial-killing case, for instance, we found four women's bodies in the same patch of woods alongside Interstate 40. The darker question underlying Dr. Bridgens' seemingly routine request was this: Could a serial killer be on the loose in northeast Kansas?

Several quick observations led me to suspect that the bones came from only one person. First, there were no duplicate bones—only one left femur, only one right humerus, and so forth. Second, all the bones appeared similarly weathered, which indicated they had probably been exposed to sunlight, heat, moisture, and cold for the same amount of time. And third, all the bones were consistent with the skeletal structure of a thirteen-year-old white female.

It was fairly easy to estimate the age. What makes growth possible for children and impossible for adults is this: During our youth, the ends of our long bones—called the epiphyses—are not yet fused to the shafts; instead they're connected to the shafts by cartilage, allowing the shafts to grow throughout childhood. Beginning in early adolescence, the epiphyses begin to fuse, gradually halting the growth of the bone shafts and bringing an end to that remarkable growth spurt occurring around puberty. On one of the long bones from Lenexa, the right humerus, the distal articular surface (the upper arm's portion of the elbow's "hinge") had fused, or united, with the shaft—a union that normally occurs around age twelve. By contrast, the medial epicondyle (a nearby epiphyses, which you can feel as a bump on the inside of the elbow) hadn't yet fused, so the humerus clearly came from someone only slightly older than twelve. The development of the teeth was also consistent with an age of thirteen or so: the second maxillary molars (nicknamed the "twelve-year molars," and with good reason) were fully developed, but the third molars (the "wisdom teeth," which normally erupt around age eighteen) had not begun to mature.

Confirming the sex of the postcranial bones wasn't as easy as you might think. During childhood, there's very little skeletal difference between male skeletons and female skeletons; it's only during puberty, when the female hips widen and the pelvic cavity deepens, that the difference becomes pronounced. Fortunately, one skeletal feature does hint at the difference, even early on: in females, the sciatic notch—the gap in the lower hip where the sciatic nerve emerges from the spine and enters the leg—begins to widen in the very earliest stages of puberty. The sciatic notch in the ilium from the field in Lenexa was the width of two of my fingers, which made it twice as wide as a male's would be. The skull had sharp-edged eye orbits and a smooth forehead, as well as small muscle markings in the occipital region—all signs of a little, lightly muscled female.

By now I was almost certain that the bones all came from Lizabeth Wilson. But sometimes, especially in murder cases, "almost" isn't quite good enough. Was there a way to make sure? I thought there might be.

My stepfather, Charlie Bass, was a geologist by training. That meant that as a child I was treated to numerous informal lectures and demonstrations highlighting the properties of various rocks and minerals, including the surprisingly dramatic glow some minerals gave off when illuminated with ultraviolet light, or "black light." In daylight, for instance, the mineral fluorite is a drab, chalky color; in a dark room under UV light, though, fluorite glows a brilliant blue; the mineral calcite shines bright red; and aragonite gives off a neon green. If you've ever stepped into a teenager's cavelike room decorated with black-light posters (less common now than they were in the 1970s, when my three sons were growing up), you've seen another version of UV fluorescence in action.

Bones are composed largely of minerals—calcium phosphate, mostly, with dashes and pinches of other minerals thrown into the mix, for reasons I don't pretend to fathom—so if you shine UV light on a bone, it will fluoresce slightly. I knew that Lizabeth Wilson's bones should all fluoresce at the same wavelength, or glow the same color. And since differences in heredity, diet, and other factors make every person's bone chemistry unique, I also knew that anyone else's bones should fluoresce at a slightly different wavelength or color than Liz's.

That, at least, was the theory. In practice, though, it wasn't quite that simple and clear-cut. I'd tried the technique in several other cases by this time, and I'd noticed that the differences in the color were not as dramatic in bone as in geologic specimens displayed in museum cases. I'd also observed slight differences among bones from the same individual. Did the decay process affect the bone chemistry slightly, and inconsistently? If some bones were protected and some exposed to the elements, would that make a difference? So while I had reason to hope that UV fluorescence could confirm that all fourteen of these bones were Lizabeth Wilson's, I couldn't be certain of that until we turned out the lights and turned on the UV lamp.

To minimize the chance that environmental factors—trace minerals in drinking water, for instance—might skew the results, I decided to compare the Lenexa bones with another Kansas case: that of a black female toddler who'd been beaten to death, then buried in a fencerow near Leavenworth. The person who killed her undoubtedly figured she'd be unlikely to be plowed up if she were buried among the osage oranges that serve as fence posts and windbreaks between farm fields. However, only a year or so after the girl was killed, the small farm where she was buried was sold and incorporated into a larger farm, so the windbreak was cut down and a bulldozer unearthed her remains—a mix of freshly fractured bones, partially healed bones, and fully healed bones: heartrending evidence of a brief life filled with bone-breaking abuse. As a further point of comparison, I brought in a third set of bones, belonging to a seventeen-year-old white female. This girl was from Tennessee, not Kansas, but she was the same race as Lizabeth Wilson, and she was closer to Liz's age.

When Doug and I scanned the Tennessee teenager's bones with the UV lamp, they all fluoresced a pale green. When we played the lamp over the Leavenworth toddler's bones, they shone a yellowish-purple. And when we trained the light on the bones from Lenexa, every one of them glowed the same deep purple. "It is our conclusion," I wrote in the report I sent Dr. Bridgens, "that the fourteen bones are all from the same individual." All from Lizabeth Wilson.

Twenty-eight years passed, and no one was ever charged with Liz's murder. Then, in early 2004, I received a flurry of calls from Senior Special Agent Brad Cordts of the Kansas Bureau of Investigation. The KBI had gotten a tip about the identity of her killer, and was about to make an arrest. They had a serious problem, though: during the nearly three decades that had elapsed since Lizabeth's bones were found and identified, Dr. Bridgens had died and the KBI wasn't sure where the bone had ended up. Brad wondered if I might have retained the bones at UT; unfortunately, I'd sent them back to Dr. Bridgens, according to the return receipt in my file, and he'd received them on February 7, 1975. I did have one piece of encouraging news for the KBI agent, though: my file included a copy of a cranial X-ray taken at UT Medical Center, while the bones were in my custody. As the last living link in the chain of skeletal evidence, I was the only person who could testify with direct authority that the remains had been positively identified as Liz's.

On September 23, 2004, I stepped off a plane in Kansas City and got into a black Lincoln Town Car that whisked me to the Johnson County courthouse in Olathe, Kansas, where I'd testified in 1971 in the case of two-year-old Lisa Silvers, after her uncle was charged with her murder. There I met with Brad Cortds, the senior KBI agent. Brad brought several items with him to the meeting: my original report from 1975; the postmortem X-ray of Lizabeth Wilson's skull; and—last but far from least—a box containing fifteen bones: the fourteen Doug Owsley and I had studied under the black light in a lab long ago, plus a fifteenth bone—a right femur—found in a hay bale a year or two after the first batch turned up. The bones had never actually been lost, as the KBI briefly feared; instead they'd been returned to the family and buried in a grave in Iowa, from which they'd recently been exhumed during preparations for the trial.

The man on trial for Lizabeth Wilson's murder was named John Henry Horton; at the time of Liz's disappearance, he was the twenty-six-year-old janitor at Shawnee Mission East High School, the area where she was last seen. According to news reports, the police and KBI had long suspected Horton, but they lacked enough evidence to charge him. That finally changed in early 2003, when a woman came forward and told investigators that when she was fourteen, Horton drugged her with chloroform and sexually molested her. One theory of the crime in Liz's death held that Horton tried the same tactic with her, but gave her a lethal overdose of the drug.

The day after I arrived and met with Agent Cordts, I was called to take the stand in John Henry Horton's murder trial. After the bailiff swore me in, Assistant District Attorney Rick Guinn led me through a series of questions designed to help the jury understand what forensic anthropology is, and what sort of qualifications I have to testify in court. Then he asked me to describe how I examined the bones back in 1975 and determined that they were all from the same young female. Finally he asked if I could tell whether the bones I'd looked at the day before were the same bones I'd studied nearly thirty years before. "They are the same bones," I answered. "Fortunately, I did a pretty good job of the record. . . . The fourteen bones are numbered, or I had numbered them, the sides and what was there, and we put them in anatomical order, and they all matched." I also described the cranial X-ray I'd taken in 1975, and explained how the air cavities that showed up in the mastoid process (the prominent bony areas just behind the ears) in the 1975 X-ray matched the air cavities in a new X-ray taken after the bones were exhumed.

Horton's defense attorney didn't challenge anything in my original analysis, nor did she question my certainty that the bones I'd examined the day before were the same bones. About all she did was ask whether I'd observed anything in the skeletal material indicating an unnatural manner of death. "There is nothing on those bones that will tell you that," I answered.

"And so probably there wouldn't be anything about those bones either that would indicate if that person had ever been molested?"

"No, ma'am."

"Nothing in the bones?"

"Not from the bones."

The cornerstones of the prosecution's case against the former school janitor included the fact that Liz was last seen near the school, where Horton was at work at the time of her disappearance, and the testimony of the woman who charged that Horton had chloroformed and molested her at age fourteen. Although the case was purely circumstantial, the jury found it convincing, voting to convict Horton of first-degree murder. He began serving a life sentence, with no chance of parole for at least fifteen years.

Then, in February 2007, the Kansas Supreme Court struck down the conviction, ruling that the district court should not have allowed testimony from the woman who claimed that Horton had chloroformed and molested her. Apart from that alleged "prior bad act," the supreme court ruling said, prosecutors lacked enough evidence even to charge Horton.

The very day the supreme court struck down the conviction and ordered Horton's release, Johnson County District Attorney Phill Kline refiled a murder charge against Horton. "It is vitally important that we ensure that Lizabeth's cry for justice be heard," Kline told reporters.

Four months after the D.A. made that statement, I stepped off another plane in Kansas, an aging scientist testifying once more about a young girl's bones he first held in his hands nearly thirty-five years ago. Will Lizabeth's cry for justice finally be heard? I don't know yet. All I know is that I've done all I can to shed a ray of light on a sad, dark crime.

4

FORENSICS AT THE SPEED OF FLIGHT: INTERPRETING TEETH AND TRAUMA

A career in forensic science isn't ideally suited to finger-drumming, foot-tapping, impatient sorts. Many cases involve weeks or even months of study, research, and meticulous drudgework by a team of methodical investigators—scientists, evidence technicians, lab techs, and detectives. Some cases resist closure for years or—like that of Elmer Reynolds—even decades. But in a few rare instances, real-life forensic progress occurs with remarkable swiftness—*almost* as fast as it does on *CSI.* In one memorable case, I managed to identify a set of scattered, fragmentary skeletal remains in the field, and to begin piecing together the brutal way in which the young victim was killed.

The key to the identification was forensic dentistry, a specialty that dates back centuries. One interesting footnote to the Revolutionary War, in fact, revolves around forensic dentistry. Paul Revere, the silversmith who became America's most famous Minuteman, was also a practicing dentist. After the Battle of Bunker Hill, he identified the body of General Earl Warren, who'd been buried in a mass grave, by means of a silver dental bridge Revere had made for the officer. Unlike DNA and fingerprints, teeth are not only distinctive but durable, as the first use of forensic dentistry in a U.S. criminal court demonstrated in 1850. Harvard chemistry professor John White Webster was accused of murdering surgeon George Parkman (to whom he owed a large sum of money) and dissecting the corpse. A set of false teeth, found in a furnace near Webster's chemistry lab, allowed Parkman's dentist to identify his body, paving the way for Webster's conviction and hanging. It was nearly half a century later, in 1898, that the first scientific textbook devoted to forensic dentistry was published, a French title that translates roughly as *Dentistry in Legal Medicine*. The first U.S. text didn't follow until the release of *Forensic Odontology* in 1966. Most fans of forensic books and television shows are aware that the biochemical building blocks of DNA can be assembled in many billions of different ways, ensuring that no two people—except for identical twins—will possess the same genetic "fingerprint."

Not many of those same people realize that there are likewise billions of possible combinations of tooth shapes, sizes, orientations, and anomalies, including cavities, fillings, chips, and distinctively shaped roots. Although identical twins can't be distinguished from one another by their DNA, they *can* be told apart by their teeth.

Trauma analysis—interpreting marks on bone to unravel how a person was killed—is far newer than forensic dentistry. My renowned mentor Dr. Wilton Krogman, who spent years teaching the intricacies of teeth, skipped right over skeletal trauma in the first edition of his groundbreaking forensic anthropology textbook, *The Human Skeleton in Forensic Medicine*, published in 1962. Surprisingly, the word "trauma" does not rate even a mention in the index of the book's second edition, published just twenty-one years ago. Another giant in the field, T. Dale Stewart, also gave short shrift to trauma in his 1979 text, *Essentials of Forensic Anthropology.*

My initial education about skeletal trauma came not from professors but from Native Americans—from examining thousands of Arikara Indian skeletons, which I excavated during the course of thirteen summers between 1957 and 1970. Among those thousands of skeletons, hundreds bore the signs of trauma: skulls gouged across the forehead and at the base with the telltale cuts of a scalping knife; other skulls crushed by war clubs; bones (most often pelvic bones) in which an arrowhead lodged for months or even years after a warrior was shot, as evidenced by the way the bone healed around the flint. Most of the Indian graves I excavated coincided with the rapid proliferation of horses throughout the Plains Indian culture, and the number of broken arms and legs soared when Indians began trying to ride horses (it's like teenagers and cars, these days). My crash course in skeletal trauma—like my immersion, at Dr. Krogman's feet, in the intricacies of teeth—was to serve me well many times during my career. But the two converged as never before, and never since, in a case twenty-five years ago.

One afternoon in March of 1982, I got a call from Dan Cook, an assistant district attorney from the Twenty-first Judicial District, a rural area located a couple hundred miles west of Knoxville. A skull had been found in rural Houston County, he said, and he hoped I could identify it. Houston County, by the way, is named in honor of Sam Houston, who was born in Virginia in 1793 but who grew up in Tennessee—a fact that probably accounts for his subsequent greatness. Over the course of a remarkable career, Houston served as governor of Tennessee, as commander of the army that defeated the Mexican general Santa Anna after the Alamo, as the first (and only) president of the Republic of Texas, as a U.S. senator from Texas, and as governor of Texas.

Unaware of Houston County's illustrious pedigree-by-association, a dog had brought a cranial vault—the top of a skull—into the yard of a trailer about eight miles east of Erin, Houston County's small (population 1,500) seat of government. The people living in the trailer had recognized the skull as human, and had called the sheriff's office, which in turn called the D.A.'s office.

Dogs have a keen sense of smell; they love to chew on bones, and with the exception of dogs that have been carefully trained to search for human remains, they don't normally make any distinction between animal bones and human bones. If someone has been killed and dumped in the woods in a rural area, it's often hunters or dogs who find the victim's body or skeleton. The favorite bones of dogs are the long bones of the arms and legs—the humerus, radius, and ulna from the arm, and the femur, tibia, and fibula from the leg—because the ends of these bones contain lots of bone marrow, which canines consider a delicacy. When dogs have found a body, they tend to pull off and scatter the bones, and they invariably chew off the ends of the long bones, turning up their noses at the bone shafts, which contain much less marrow.

Skulls tend to pose a problem for most dogs. The skull is big and round, like a bowling ball—too big to fit easily into a dog's mouth. The mandible, or lower jaw, and the cheekbones, or zygomatic arches, offer projections a dog's teeth can clamp down on, but once those have been chewed off, it takes a pretty big dog to latch onto a skull and bring it home, and by the time Fido trots up the front steps with it, there's not much left of the bones of the face.

Assistant District Attorney Cook asked if he could send the skull to me in hopes I might be able to tell whether it could belong to a sixteen-year-old girl from a neighboring county who had been missing since the previous November. The girl, a pretty honor student who kept up her grades while holding down a part-time job as a waitress at a steakhouse, had vanished after dropping her boyfriend at his house. The next day, her locked car was found in a church parking lot just two blocks from her home.

The missing girl's name was Kathy Nishiyama, and when I asked Cook whether one or both of her parents were Japanese, he told me that her father was. At that point, I figured there was a pretty good chance I could tell, just by looking at the skull, whether it was likely to be hers.

Japanese people are classified by physical anthropologists as "Mongoloid" (descended from ancestors in Mongolia), with skeletal features that distinguish them from the other major groups or races, "Caucasoid" (white European) and "Negroid" (black, of African descent). In addition to including most Asians, the Mongoloid group includes Native Americans, who are descended from Asians who crossed the Bering Strait many thousands of years ago and migrated down through North America and South America. One hallmark of Mongoloid skulls is their wide, flat cheekbones; another is the distinctive shape of the central teeth, or incisors: the back side of the incisors is scooped, or concave, rather than flat; if you look at them from the biting edge, the cross section resembles that of a garden spade or an old-fashioned coal scoop. For this reason, they're called "shovel-shaped incisors." I can't explain why Mongoloid peoples evolved this particular dental characteristic many thousands of years ago—as far as I know, nobody has yet figured out the "survival value" of shovel-shaped incisors to people in ancient Asia—but their forensic value is quite high, since they can reveal a murder victim's race even if virtually no other skeletal evidence remains intact. I asked the prosecutor whether any of the incisors were still attached to the skull. Unfortunately, they were gone—the mandible and even the upper jaw had been chewed off, leaving only the cranial vault—but I still hoped I'd be able to tell whether the skull was Mongoloid.

Cook arranged for Mike Dover, the chief helicopter pilot for the Tennessee Highway Patrol, to fly the skull to me that same afternoon. We hung up the phone around 1 P.M.; three hours later, a big blue and white Huey helicopter bearing the official seal of the THP on either side settled onto the helipad of the UT Medical Center in a blast of rotor wash that buffeted me and two of my graduate assistants, Steve Symes and Pat Willey.

Dover handed me the skull and I cradled the top of the cranial vault in my hands. The zygomatic arches, or cheekbones, had been bitten away, making it difficult to tell if this person had the wide, flat cheekbones typical of Asian people. Still, I could tell a couple of important things from the vault: the small size, gracile (smooth) shape, and sharp-edged eye orbits told me it was female, and the prominent cranial sutures—some of which hadn't yet begun to fuse—told me it was an adolescent. Nothing, in other words, excluded the possibility that this was a sixteen-year-old Asian-American. So far, so good, if "good" can ever be applied to the quest to identify a missing and possibly murdered girl.

The cranial vault wasn't all Dover brought me. He also brought news that had been radioed to him while he was in flight. Down an old logging road near the trailer where the skull-toting dog lived, a team of searchers had found more bones and some clothing. Could I fly back to Houston County with Dover, the authorities wanted to know, and help identify the additional skeletal material? I agreed, and offered to bring Steve with me. I was wearing a suit—hardly the best clothes for crawling around in the woods at night—but Dr. Bob Lash, the Knox County medical examiner (whose morgue was located in the hospital basement), offered to loan me a jumpsuit, and within a few more minutes we were airborne.

By the time the helicopter reached Houston County, darkness was falling. In the twilight, from a couple thousand feet above the wooded hills, Dover was having difficulty pinpointing the search area and, more to the point, his landing zone. He got on the radio and made a request, and a few moments later the dusk was pierced by the blue strobes of half a dozen police vehicles, arrayed around a small clearing. We landed and got out, and the search team led us to several evidence flags planted in the forest floor, marking the locations of several pieces of skeletal material and clothing they'd found scattered across several hundred yards in the woods.

The most significant thing they'd found was the upper jaw, which contained thirteen teeth, some intact, others broken. I immediately zeroed in on the incisors. Of the four upper incisors—the two central incisors and the two lateral ones flanking them—two were broken off at the roots and missing their crowns. Two, however—the right lateral incisor and the left central incisor—gave me the information I needed: the incisors were shovel-shaped—not dramatically, but about what I'd have expected in the child of one Mongoloid parent and one Caucasoid. The odds that this was Kathy Nishiyama had just skyrocketed.

"You know," I said to Dan Cook, "if we could get in touch with Kathy Nishiyama's dentist and get hold of her dental records, I bet we could make a positive identification this evening while we're out here at the scene." A few minutes later, as Steve Symes and I sifted through the leaves on the forest floor, I heard a turbine spool up and the helicopter ratchet skyward. Kathy Nishiyama's dentist had been located in her hometown of Clarksville, about twenty miles east, and Dover was dashing over to retrieve her dental records.

The clothing found scattered in the woods consisted of a pair of bloodstained blue jeans, which had been torn or cut; a purple sweater; a white coat; and a pair of blue and white tennis shoes. The additional skeletal material was two fragments of the right temporal bone, from just above the ear; the mid shaft of the left fibula, or shinbone; and the crown of a tooth—an upper left central incisor, whose curving posterior cross section resembled that of a shovel . . . and whose broken base fit perfectly with the corresponding root that remained embedded in the jawbone.

The police and sheriff's officers had set up portable work lights in the woods, which illuminated the areas where the additional skeletal material had been found during the afternoon's search. But the area over which the bones and clothing had been scattered was so large that a detailed search would be nearly impossible in the dark. The D.A., sheriff, and police investigators agreed to resume the search the next morning.

But our work wasn't done for the night. To the east, we heard the thumping of helicopter rotor blades, and soon the landing light of Mike Dover's chopper swooped into view. In less than an hour Dover had flown to Clarksville, retrieved Kathy Nishiyama's dental records, and returned. It took only a moment to make the comparison. I'd already noticed, in my initial examination of the skull, that the teeth were in rather poor condition, and not just because several of them had been snapped off at the gum line. Of the thirteen teeth in the upper jaw, ten had fillings, and one had three fillings. In addition, one of the left molars had been pulled before death, long enough ago that the root socket had already begun to fill with bone. Cavity for cavity, filling for filling, the teeth from the skull matched the dental record. Within an hour after I had stepped out of the THP helicopter, and while we were still at the death scene, we had a positive identification of previously unidentified skeletal remains: swiftness I'd never experienced before, nor since.

By 11 P.M. that night, Steve and I were gazing out of the bubble of Mike Dover's cockpit once more, admiring the lights of Nashville, spread out like jewels on black velvet beneath us. It was one of the prettiest sights I've ever seen, and contrasted sharply with the ugliness of the death I'd been asked to help solve. Dover touched down in Nashville just long enough to refuel; by 2 A.M., I was back in Knoxville and asleep in my own bed. While I slept in my bed, Kathy Nishiyama's skull rested fifty feet away—though "rested" seems the wrong word—in my locked car in my locked garage.

Identifying Kathy Nishiyama was only the first step. Determining what had killed her—and who—were the crucial next steps. I'd brought her skull back so I could study it closely, to be certain of her manner of death. But I'd seen enough already, since the moment Mike Dover first handed me the cranial vault at the hospital helipad, to form a fairly clear and horrifying picture.

The broken teeth—three of the four incisors were broken off near the root—offered one important clue. They exhibited what's known as a "hinge fracture": the front surface of each tooth had snapped completely across in a clean horizontal line, and the teeth had folded backward into the mouth like a hinge flexing; the hinge fractures meant they had been hit from the front with great force. The blow carried so much force, in fact, that not only did the teeth snap backward, but as they did, their roots acted as levers, bursting through the bones on the front (anterior) surface of the jaws. The right lateral incisor was the only one of the four central teeth that wasn't snapped off, but it was chipped on its medial surface (the corner closer to the midline of the jaw). That meant the blow probably came from slightly to Kathy's left, as it most likely would have if a right-handed person were swinging an implement horizontally. This blow, while certainly painful, would not have killed her, and might not even have knocked her unconscious.

Another horizontal blow caught her on the forehead, above the left eye. This one left an oval indentation in the outer table (layer) of the skull, but did not strike with enough force to fracture the innermost of the skull's three layers. The cross section of the depressed fracture measured an inch long (horizontally) by three-fifths of an inch high. The oval shape could have been produced by a blunt implement that was cylindrical, or round in cross section: something like a baseball bat, tire iron, or four-cell flashlight.

Another, weaker blow caught her almost in the midline of her forehead, just above the eyebrows. This one also left an oval depressed fracture, oriented vertically, that was about half the size of the one over the left eye. Like the larger oval, this one did not fracture the inner table of the cranium.

None of these three blows, in my opinion, was powerful enough to have killed Kathy Nishiyama, though either of the two frontal fractures could certainly have knocked her unconscious. Then came a fourth blow: an impact behind her right ear that was so powerful it completely sheared off the rear portion of the temporal bone in two pieces. (Those pieces were among the additional skeletal material recovered while Mike Dover was flying eastward to pick me up in Knoxville.) Besides shattering the temporal bone, this blow created a fracture more than three inches long, extending through both the outer and inner tables of the skull. The geometry of these fractures differed from the others. As best I could tell, this blow came while she was lying on the ground, with the right side of her head upturned. This blow could have come from a baseball bat or tire iron or flashlight; it could also have been inflicted, I thought, by a heavy boot, stomping down on the side of the girl's head as she lay on the ground. There was no way to tell, but I could only hope this poor child had lost consciousness early in her ordeal, so she didn't suffer for long.

It didn't take the police long to identify a suspect. On Tuesday—just one day after her skull was found, and while I was diagramming the fractures on an outline drawing of the skull—investigators were questioning a local man.

The suspect, twenty-three-year-old Eddie Hartman, should have had an ironclad alibi: Hartman was an inmate of the Dickson County jail, serving a three-year sentence for burglary, at the time of Kathy Nishiyama's disappearance. But while Hartman was an inmate *of* the jail, he wasn't an inmate *in* the jail. Not, at least, the night she vanished. Hartman was a "trusty"—one of your higher-ranking convicts, whose good behavior and dependable nature had earned him special privileges. One of those privileges was working outdoors in the fresh air, on a tobacco farm owned by a Dickson County sheriff's deputy, Sergeant Carroll Fizer. On November 16, 1981, the day Kathy disappeared, Hartman worked at Fizer's farm until dark. Rather than drive the prisoner back to jail in town, the deputy gave Hartman the keys to his cruiser and told him to drive himself. With a population of just 800, Charlotte was a small, sleepy town, a sort of Middle Tennessee version of Mayberry; the deputy may even have been thinking of how Otis, the lovable town drunk in *Andy Griffith* episodes, would obligingly lock himself up whenever he needed to sleep one off. Except Hartman, who had been convicted of abducting a teenaged girl in 1978 (she escaped by jumping out of his car), was no lovable lush, and this was no warmhearted sitcom with a happy ending.

Instead of driving straight back to jail, Hartman drove to Clarksville, whose 60,000 inhabitants made it a teeming metropolis in comparison to sleepy little Charlotte. He didn't return to the jail until 3:30 A.M., and when he did, he seemed "real scared," the Dickson County sheriff later said. Something resembling spattered blood was seen on the patrol car's right-rear fender and trunk by four officers later that day, too.

When Kathy Nishiyama's remains were found, people remembered Eddie Hartman's evening of freedom, his late return, his nervousness, and the stains on the car. Two witnesses came forward to say that on the night of November 16, they'd been pulled over near Clarksville—in Montgomery County—by a man in civilian clothes who was driving a Dickson County sheriff's cruiser. Another witness said he saw a patrol car, its lights flashing, beside a car in the church parking lot where Kathy Nishiyama's car was found.

Three and a half months after Kathy Nishiyama's disappearance, the patrol car was impounded and turned over to the Tennessee Bureau of Investigation to be searched for evidence—not exactly ideal working conditions for forensic technicians. Despite the delay, their search yielded at least one noteworthy find: behind the car's rear seat cushion they found a gold necklace, which Mrs. Nishiyama thought she recognized as Kathy's. They also found a tire iron and a large police flashlight, both of which a TBI agent was able to verify were in the vehicle on the night of November 16—and either of which could have caused the fractures in her skull.

Fourteen months after a dog brought a bludgeoned cranial vault into the yard of a rural Tennessee trailer, Eddie Hartman stood trial for kidnapping and first-degree murder. In addition to my testimony about the identification and the skull trauma, the prosecution produced a string of witnesses who testified to seeing Hartman in the patrol car in and around Clarksville on the night of Kathy's disappearance. The testimony of one other witness proved especially damaging to Hartman's case—a prison inmate who said that Hartman had described luring Kathy into the patrol car with a bogus story about a serious illness in her family. After that, the inmate said, Hartman detailed how he raped her in the cruiser's backseat, killed her, and then sexually abused her body.

Toward the end of the trial, Hartman's attorney described the lead prosecutor as "a desperate lawyer grasping at a straw," then added, "That straw is out of reach." The jury did not agree with him: the twelve jurors deliberated for just forty-five minutes before finding Hartman guilty on both counts. He was sentenced to death. When he appealed the death sentence, I testified again; as I described the skeletal trauma to the jury, my wife Carol kept her eyes on Eddie Hartman. (Carol is my third wife, whom I married in 1998. My first wife, Ann, died of cancer in 1993; my second wife, Annette, also died of cancer, in 1997.) Carol later told me that as I described each injury and how I thought it had been inflicted, she saw Hartman nod repeatedly.

For years Eddie Hartman continued to appeal his death sentence, though as far as I know, Kathy Nishiyama never got the chance to appeal hers. On May 24, 2007, Hartman—age 49—died in a Nashville hospital of natural causes, according to a prison spokeswoman. In a way, it's as if cosmic justice stepped in to carry out the death sentence Hartman had been avoiding. But then again, none of us gets out of here alive.

5

THE ROCKETS' RED GLARE, BODIES BURSTING IN AIR: DEALING WITH A MASS DISASTER

Identifying one body can be a daunting challenge; identifying a dozen—and having to find their pieces and reassemble them first—can be overwhelming. It's the sort of nightmarish scenario that mass-fatality teams now train for, against the possibility of airliner crashes and terrorist attacks. And it's the scenario I suddenly found myself confronting on a warm Friday in May nearly twenty-five years ago, in a part of Tennessee so rural and sleepy, I would never have imagined a small-scale apocalypse unfolding there, requiring swift, coordinated response by half a dozen local, state, and federal agencies.

Shortly after noon on May 27, 1983, I was just finishing lunch at the UT Faculty Club. The spring semester had ended, and summer classes had not yet begun, so I was winding down for a quiet weekend. Or so I thought.

Then Annette, my secretary, called from her office under the football stadium and had me summoned to the phone. "TEMA just called," she said. Not a good sign. TEMA was the Tennessee Emergency Management Agency, and if TEMA had a problem, it was likely to be a big one. "They want you to call this number in Nashville." I took down the number and called.

I didn't get much information from the call, but what little I got confirmed my worries. There had been an explosion in nearby Polk County, the TEMA official on the other end of the line said; they needed my help identifying an unknown number of mangled bodies. They had already dispatched a plane from Nashville to Knoxville to pick me up. The explosion had occurred around eleven-thirty; they reached me at one; the airplane would be touching down in half an hour.

I quickly rounded up two of my best graduate students, Bill Rodriguez and Steve Symes. Both were well trained in osteology—how to identify bones, and even fragments of bone. In addition, Bill had conducted a pioneering study of insect activity in human corpses, and Steve was on his way to becoming an expert in bone trauma (specifically, cut marks in bone). The three of us had just arrived at the general-aviation terminal at the Knoxville airport when the plane landed. Less than an hour after my lunch was interrupted, Bill, Steve, and I were airborne in a twin-engine state plane. We had two more passengers to pick up on our way to the site: Dr. Cleland Blake, a forensic pathologist, and his assistant, Lane Moore. Blake and Moore boarded the plane at the small airport in Morristown—about forty miles northeast from Knoxville, as the crow flies—and then we turned southeast, skirting the western flank of the Great Smoky Mountains and then the lower foothills of the Appalachians.

Polk County lies roughly a hundred miles below Knoxville, in the extreme southeastern corner of Tennessee. Bordered on the south by Georgia and on the east by North Carolina, much of the county is mountainous, and very little of it is populated. Back during the Civil War, Copper Hill was a booming mine town that provided 90 percent of the copper used by the Army of the Confederacy. Between the mining itself and the acid runoff resulting from it, Copper Hill came to resemble a moonscape: devoid of vegetation, its barren contours a bright, unnatural-looking orange. The mines began closing down in the 1970s, though, and in recent decades the focus in Copper Hill has shifted to restoration: healing a landscape that was laid waste by mining, erosion, and toxic chemicals.

Apart from the Copper Hill Basin, much of Polk County is farmland and forest, including 50,000 acres of the Cherokee National Forest. Two beautiful rivers flow out of the mountains within the forest: the Ocoee, whose pounding class IV and V rapids attract hardcore kayakers and thrill-seeking rafters; and the Hiwassee, a gentler river whose class I, II, and III rapids draw people in canoes, rafts, and even inner tubes.

Despite the stop in Morristown, we still touched down in Cleveland by 2:30 P.M. State troopers met us at the airport and drove us east to Benton, the town of about 1,000 people that serves as the county seat of Polk County. We took a two-lane blacktop south from Benton. As the highway patrol cruisers careened around the winding road at 80 miles an hour, it occurred to me that there might soon be a few more casualties before the sun set. I cleared my throat and said to the trooper at the wheel, "You know, these folks are already dead; they're not going to get any deader if we slow down a little bit." He didn't take the hint, and we continued to rocket along. Within a few miles I saw fifteen or twenty vehicles—ranging from police cruisers and ambulances to sightseers' pickup trucks—lining the roadside.

A large sign at the edge of the road identified the property as Webb's Bait Farm, offering red worms and fishing tackle for sale. The owner, I learned, was a thirty-year-old man named Dan Webb. Dan Webb raised happy worms, it appeared: a big, grinning worm dominated the sign, wearing a floppy fishing hat and carrying a rod and reel over what would be his shoulder, if he'd had a shoulder; dangling from a line in his right hand—he actually did have hands, this remarkable worm—was an enormous fish. Atop one corner of the sign, up near the worm's head, was a video surveillance camera, aimed at the driveway. Not the sort of thing I'd have expected to see at a worm farm, given that I'd never heard of a worm-farm robbery. But then again, I'd never heard of a worm-farm explosion, either, so perhaps there was a lot I didn't know about worm farms.

As we parked along the shoulder, joining the fleet of other vehicles, I noticed wreckage scattered across a hundred yards or more of hillside—cleared land, mostly, bordered on the south by a stand of young pines. There was a gap in the tree line; a closer look revealed that trees littered the ground there, felled by the force of the blast. Between me and the fallen trees, about fifty yards from the smiling worm, lay a lumpy white shape, which I recognized at once as a body covered by a body bag. A hundred yards beyond that was a trail of splintered lumber and twisted roofing tin, beginning at a shattered foundation, running toward the gap in the trees, and continuing deep into the woods. Only a few hours before, a large barn had occupied the foundation; now, it looked as if a ferocious tornado had touched down and decimated, with pinpoint precision, a single unlucky structure. As soon as I saw the devastation—and heard that a dozen or more people were thought to have been working in the barn when the explosion occurred—I knew that Steve, Bill, and I had our work cut out for us.

At first the cause of the explosion was a mystery, and the locals were reluctant to answer questions, but before long the truth emerged. Webb's Bait Farm, it turned out, produced more than just worms; it also produced fireworks, illegally and—obviously—dangerously. Although the factory was set in a rural area and all the workers were related to Dan Web by blood or by marriage—including his wife, his mother, his brother, and his uncle—this was no small-time mom-and-pop operation; it was the largest illegal fireworks factory ever found (if "found" is the right word for it) in the United States, churning out millions of M-80 and M-100 firecrackers (which contain the equivalent of a quarter stick of dynamite apiece).

The main part of the operation was housed in the barn, or what used to be the barn. To one side of the barn, several dozen grave-sized enclosures stood knee-high off the ground: the rectangular beds that produced the operation's legal product, fishing worms. A stone's throw from one end of the barn's foundation, a white house still stood. A long, skinny building, it had been cobbled together over time; alternating regions of horizontal clapboards and vertical board-and-batten siding suggested at least three additions to the tiny original structure. One end was stripped of siding, and the long rear wall was buckled, but the house had fared remarkably well, considering how efficiently the nearby barn had been rendered to matchwood.

A two-door Toyota Corolla, parked near the barn, had also sustained some damage but largely survived. Its windows and rear windshield had shattered, and the right front quarter panel was gone. So were all four hubcaps, exposing shiny hubs underneath. The trunk lid had been bent upward by a couple of inches, and the car's sides bore countless scrapes from flying debris. The skin of the passenger-side door—which was facing the barn—had absorbed enough of the blast to flex inward, revealing the outlines of three horizontal steel beams in the door's core.

TEMA wasn't the only agency, and not even the ranking agency, at the scene. Overall responsibility lay with the federal Bureau of Alcohol, Tobacco, and Firearms (ATF), which enforces federal laws governing explosives. The Polk County Sheriff's Office was out in force, controlling access to the scene. The Tennessee Bureau of Investigation was coordinating the recovery and identification of people killed in the explosion, which is where Dr. Blake and I came in, along with our assistants.

I was anxious to get started. We had a lot of bodies to recover, and on a hot afternoon like this, decomposition and blowfly activity start fast, and gain momentum with dismaying speed. I told one of the TBI agents we'd need some refrigeration, and he radioed in a request for two refrigerated semitrailers, which—amazingly—arrived in a matter of hours.

But before I could start my work, a search warrant had to be served. When the barn blew up, Linda Sue Webb was reportedly in the house, but by the time the first sheriff's deputy arrived, she had fled. Her husband, Dan Lee Webb, was in New York on business—selling fireworks, I supposed, to dealers stocking up for the Fourth of July. A deputy district attorney arrived around four o'clock to serve the warrant; lacking the presence of either owner, he served the warrant on the house itself, reading aloud as if the battered structure could hear. It was a bizarre legal ritual I had never before witnessed, though I've seen and heard it a few times since.

But even after the warrant was served, we weren't yet free to search for victims. One big worry of the ATF agents was the potential danger from "unexploded ordnance": fireworks that had escaped the initial blast. And, indeed, a mobile home parked near the wreckage of the barn—one side of it peeled open like a sardine can's lid—was jammed with enough firecrackers and bottle rockets and roman candles to shred the trailer and anybody near it. Luckily, the mammoth explosion—some witnesses claimed to have seen a mushroom cloud—had consumed the oxygen around the barn, so the blast itself had snuffed out the sparks and embers that might have set off the stockpiled fireworks. Just to be on the safe side, though, the ATF agents oversaw the bulldozing and burial of the trailer's contents beneath several feet of red clay.

The bulldozing wasn't done until 4 P.M.—later than I'd hoped—but with the summer solstice only a few weeks away, we'd still have a good four hours of daylight to find the dead and begin piecing them back together. We would work in stages, I decided: first we'd do a quick search for bodies or torsos, so we'd know how many victims there were; then we'd go back and do a more detailed search for missing hands, feet, arms, legs, and other parts.

Miraculously a teenaged boy working in the barn had survived the blast. He, like almost everyone else inside, had been blown through the roof. His trajectory carried him over the house, and he landed in the front yard. When the first emergency personnel arrived, he'd been wandering around the yard, shell-shocked. He'd already been taken to the hospital by the time I arrived, but as I began surveying the carnage, I marveled at his luck.

His good fortune was underscored by the fate of body number 1, a large woman who had also been blown over the house. She hit the carport roof, punched through the corrugated fiberglass as if it were tissue paper, then skidded across the concrete, her path marked by a wide streak of grease. Shreds of fabric clung to her shoulders and arms; the rest of her body—such as it was—was nude, her clothing ripped away by the force of the blast and her buttocks turned silvery-gray, the color of gunpowder. She lay facedown, her upper body on the edge of the concrete, her lower body on a piece of the roofing she had shattered. Her left leg looked like a pirate's peg leg: the foot had been blown off, and all the flesh stripped from the bones of the lower leg. Her right leg was largely intact; the shoe had been torn from that foot, but it had stayed on long enough to protect the sole of her foot: her toes were undamaged, though her ankle was shredded. The worst damage was at the other end of her body, though; the occipital bone—the base of the skull—was there, but the rest of her head was simply gone.

Body number 2 had crashed through the metal roof of the house itself, splintered the two-by-ten-inch rafters, and come to rest in the attic. Another large white female, she, too, had been stripped by the blast. A two-inch drywall screw lay beside her; a piece of foamboard insulation was beneath her, while particles of the white foam and splinters of rafter were sprinkled atop her seared, bloody torso. She was lying on her stomach, her right arm tucked beneath her and across her body, emerging near her left shoulder. Her left shoulder did not have an arm attached to it; the arm had been sheared off about two inches below the head of the humerus. Large smears of blood and tissue marked the spot where the woman's head had impacted. The top of her head was missing, and her brain lay four feet away, next to a Christmas video—*Here Comes Santa*—whose box featured a cartoon train smiling at me from beside the brain.

Body number 3—yet another nude female—was the only one that was not hurled from the barn by the blast. She was lying faceup, her arms flung back over her head as if she'd been sitting near the origin of the explosion;

as if, when it happened, it toppled her backward. Her flesh was burned to a crisp grayish-black, brittle and crumbly. Her lower legs were both missing from mid-shin down; her forearms and hands were gone, too.

By now the refrigerated trailers had arrived, and since many of the bodies were disarticulated, or blown apart, I decided to designate one trailer for torsos, the other for appendages. We had no tables to work on, so we covered the floor of each trailer with plastic tarps to keep the wood from getting soaked with blood.

Police and other emergency personnel who frequently rub shoulders with death will understand this next bit of foolishness; others may be startled to read of a practical joke at a mass-fatality scene. Bill Rodriguez had gone through the woods looking for bodies; to get back to the command post more easily, he came out of the woods, climbed over the fence and the edge of the pasture, and walked back along the road. But when he got to the driveway, the deputy securing the entrance wouldn't let him in. Bill saw Steve Symes nearby and called to Steve, "Hey, tell this guy I'm part of the forensic team." Steve assumed his most solemn expression and assured the deputy that he had never seen Bill before in his life. After several minutes of confusion, I had to come out and vouch for Bill. I told Steve that was not the sort of stunt to pull at a crime scene, but the truth is, the interruption gave us a few moments' relief from the horrors strewn across the property.

Out in the pasture, between the barn and the woods, a line of flattened grass and a gouge in the earth marked the place where a body had hit, then skipped another fifteen feet before coming to rest. This one was a young, slim white female—a nineteen-year-old girl, we learned, who had just graduated from high school a week before and was working to earn money for a trip to Europe. All the deaths were a shame and a waste, of course, but hers struck me as particularly poignant. Unlike the older women, she was not nude; she was wearing blue jeans, and denim is pretty tough. The right leg of the jeans showed little damage, but the left pants leg had shredded, and the concussion had split open the left thigh. Her shirt had been peeled up over her head, her arms still caught in the sleeves; her bra was undamaged. Her legs, back, and shoulders were mottled with patches of pale, almost white skin, where the searing heat had blanched the flesh; I knew from my research that victims of mine explosions also tended to have blanched skin. Sadly, her mother was also one of the victims.

As my students and I worked to find bodies and body parts, the arson investigators and explosives experts were combing through the debris in search of what caused the explosion. Eventually they found the charred remnants of an electric drill, with a mangled paint-stirring attachment still cinched in the drill's chuck. Their theory was that one of the workers was using the drill to mix a slurry of explosive ingredients when a spark from the drill's motor—and an electric drill produces a lot of sparks, as you know if you've ever used one in the dark—touched off the mixture.

At the edge of the woods we found the guy who might have been doing the mixing. He was lying facedown—or, at least, his upper body was—but his torso had been twisted 180 spine-snapping degrees, so his pelvis was facing up. That put his right leg where his left leg should have been; his left leg was nowhere near his body. His abdomen had been ripped away; some of his entrails were exposed, and the rest were missing. So was all but the top few inches of his left arm. Surprisingly, his face was relatively undamaged. Perhaps he had looked away—maybe stealing a glance at the attractive nineteen-year-old—as the fateful spark shot downward and the world exploded around them all.

The variations on the theme of catastrophic force were striking. Body number 10, one of the three male bodies, was nude—except for his leather boots, still tightly laced around his feet and ankles. When we took off the boots, we saw that his feet were purple with bruises. One of his hands lay detached but nearby in the grass. His head was nowhere in sight.

Body number 11 was perhaps the most bizarre of all. The explosion hurled him the farthest—nearly a hundred yards, skipping across a pasture and into the woods. As he entered the tree line, he snapped a branch on a pine sapling. The jagged end of the branch pierced his abdomen and snagged a loop of his small intestine, and as the body continued into the woods, the intestines unspooled like fishing line. Once he reached the end of the line, so to speak, he was yanked to a stop. We found him there—that is to say, we found his torso—at the end of twenty feet of intestine, still stretched tight as a rope.

People were not the only casualties. In the grass of the pasture lay a rabbit; it had been killed by the blast's shock wave, but its fur was completely unmarred. Closer to the barn was a dead chicken, its tail feathers gone, showing it had been facing away from the blast. Another chicken had survived, despite the loss of one leg, but it remained rooted in one spot, trembling for hours, until finally one of the deputies put it out of its misery.

By the time the sun set on the scene of destruction, we had found eleven bodies or torsos, along with dozens of arms, legs, hands, and feet. We locked the trailers for the night, not that anyone in their right mind would have been tempted to steal their contents, then drove from Benton to the nearby town of Cleveland. You might think the day's horrific sights (and smells) would have killed my appetite and made sleep impossible, but the truth was, I was starving and exhausted. I ate like a pig and then slept like a baby.

Saturday morning we began by making another sweep of the property, this time looking in more detail for smaller body parts—combing through shattered boards, looking under pieces of roofing, even scanning the trees overhead. My oldest son, Charlie—a high school teacher with a master's degree in anthropology—came up from Atlanta to help us search. We did find additional pieces—a few ears and teeth and skull fragments amid the debris; a mandible and a foot-long section of spine (half a dozen vertebrae) in the woods—but after a while it became clear that we'd recovered everything that was reasonably recoverable.

It was time to piece the dead back together again, determine their sex, and identify them. But where to begin? We clambered up into one of the chilly trailers and surveyed the mangled torsos.

Normally, when a forensic anthropologist is confronted with an unknown corpse or skeleton, the first step—before attempting to identify the victim—is to answer four key questions: What's the race? What's the sex? What's the stature? and What's the age? In this case, though, we were trying to match victims with a list of names the police had assembled from relatives, neighbors, and onlookers who had known some of the victims. That would simplify the task of identifying the bodies considerably. The bigger challenge would be reassembling them correctly. Although many of the bodies were disarticulated and battered, it was still easy to distinguish the three male torsos from the eight females.

Next came the tougher problem of matching a daunting stack of severed arms and legs to the proper torsos. We would start, I decided, with legs. To simplify the task, we divided the legs into lefts and rights, then further subdivided those into shaved and unshaved legs. That way, if a female torso (and the majority of the dead were women) was missing a left leg, for instance, we could go to the pile of shaved left legs and look for one whose break or length or girth matched the torso's. With no tables to work on, we were simply crouching and slipping around on the bloody plastic lining the cold trailers.

The work went surprisingly quickly. Within a few hours we had paired all the loose parts with torsos, and

I felt certain that even without DNA testing (which wasn't yet available back then), we had matched things up correctly. We hadn't found everything—a few fingers and toes and ears eluded us, and substantial portions of several faces and skulls were simply decimated by the blast—but we found most of almost everyone, which was remarkable, considering the force of the explosion.

In a community as small as Benton, giving names to the dead proved simple in the end. Although the locals had been reluctant to tell the authorities about the illegal activities going on at Webb's Bait Farm, they were extremely helpful in identifying victims of the explosion—victims who, after all, were locals, too: neighbors and friends and family members.

By Saturday afternoon, with all the dead reassembled and identified, my work was done. We squeezed into the cab of a UT pickup truck for the drive back to Knoxville. It was slower than my flight down, but that seemed fitting, given the widespread mourning we were leaving in our wake.

A month after the explosion, Dan Lee Webb was charged by the state with eleven counts of manslaughter, one for each death, as well as one count each (by the federal government) of manufacturing and possessing illegal explosives. He ended up pleading guilty to the manslaughter charges and received a ten-year sentence, which he served concurrently with a ten-year federal sentence for possessing and making illegal explosives. He also received a much tougher sentence for life: the lifelong burden of knowing that his dangerous and illegal enterprise had killed his mother, his brother, and his uncle.

If a similar disaster occurred today, DMORT—the Disaster Mortuary Operational Response Teams—would swing into action, arriving on the scene with portable morgues and a multidisciplinary team of forensic anthropologists, dentists, DNA technicians, pathologists, counselors, embalmers, and other professionals. DMORT is administered by the U.S. Department of Homeland Security, but most of its specialists—including a number of my former students, colleagues, and friends—are volunteers. DMORT volunteers spent weeks helping to find and identify the dead in New York after the World Trade Center attacks, and weeks doing the same thing in Louisiana after Hurricane Katrina. Back in 1983 DMORT did not yet exist, but TEMA, the TBI, the ATF, and other state and local agencies did a good job of pulling together a rapid and effective response to what counted, in rural Polk County, as a mass disaster.

Six weeks after my hasty flight south, Independence Day arrived. Skyrockets bloomed and boomed against the night sky of Knoxville, to the cheers of thousands of spectators. But I couldn't help thinking of a dozen people launched heavenward by a stray spark a hundred miles to the south.

6

DEAD FOR THE HOLIDAYS: DETERMINING TIME SINCE DEATH

Prosecutors, like forensic anthropologists, are often involved with murders. Seldom, though, are they playing the part of the homicide victim. One notable exception I encountered was Assistant District Attorney Ted Barnett. On many occasions I have estimated time since death (TSD) *for* a prosecutor. In Barnett's case, I was asked to estimate TSD *of* a prosecutor.

Barnett, age thirty-one, was prominent both in legal circles and in Knoxville's African-American community. His father was a math teacher at Austin East High School, where Ted had made good grades and had excelled at basketball. After law school, he'd worked briefly in the public defender's office in Nashville, but he returned to Knoxville to be close to his family. He bought a house in the College Hills neighborhood—adjacent to Knoxville College, where he often played basketball with the students.

In December of 1987, Barnett took two weeks of vacation before Christmas. When he failed to show up for work the following Monday, December 28, his boss, Attorney General Ed Dossett, got worried. That's not surprising, since Barnett's job—putting criminals in jail—would naturally tend to make enemies among some of Knoxville's less savory citizens. By the second day, with no word from Barnett and no answer to repeated phone calls, the D.A. and his staff were truly alarmed. Barnett's colleagues kept trying to phone him; they also went to his house and knocked on the door. Barnett's car wasn't there, however, so they weren't surprised when he didn't answer. They left notes at the house asking him to call as soon as he returned.

By the third day, they were sufficiently concerned to take more drastic steps. According to the story I heard, the district attorney himself borrowed a ladder from a neighbor, climbed in an open second-story window, and reappeared at the front door, letting in several of his staff. I can only assume that he realized the irony—Knox County's chief prosecutor turning second-story man, and making accomplices of his employees; I also assume the group's genuine concern for Barnett's safety overcame any legal concerns they might have had. (Newspaper accounts later said they had received permission from Barnett's father, who had also been trying unsuccessfully to contact Ted, so while the D.A. didn't go so far as to obtain a search warrant, he wasn't exactly breaking and entering, either.)

In a quick search of the house, the D.A. and his staff found nothing to indicate foul play. The place was messy, with evidence of some partying—there were empty wine and liquor bottles, and what appeared to be traces of semen on the bed—but there was no bloody body in the sheets, or facedown on the living room floor, or slumped at the kitchen table.

The next day, Dossett and his staff went back to the house. This time they took it slowly, and they opened doors they hadn't bothered to look behind during their first, hasty search. One door, which appeared to lead to a closet, actually opened onto a small bathroom, tucked into a shedlike addition at the back of the house. There on the bathroom floor, his head near the base of the commode and his feet in the shower, lay the district attorney's young assistant prosecutor.

Shortly after that, my friend Art Bohanan—a senior criminalist with the Knoxville Police Department—received an urgent message from KPD headquarters. I didn't learn of Art's involvement with the case until much later, but the details remained vivid in his memory when he was free to tell me. "Call the D.A.'s office ASAP," the dispatcher told Art. "They asked specifically for you."

Art called the D.A.'s office and spoke with Bob Jolley, another assistant district attorney. Jolley explained the reason for the urgent message and asked Art to come to Barnett's address right away. "Don't tell anybody," he added. "The chief is the only other person at KPD who knows about this." Art raced to the scene, where Barnett lay just as Dossett and his staff had found him. The body was bound, the hands and feet tied up with television cable, and was covered with stab wounds, dozens of them. As he studied the body, Art noticed that some of the wounds were beginning to pucker at the edges—early stages of healing—and he realized with horror that Barnett had been killed slowly, perhaps over the course of several agonizing days.

As soon as the body had been photographed, examined, and taken to the morgue, Art set to work, seeking fingerprints and other evidence. He found plenty. Besides not finding the body in their initial search, the D.A.'s office had overlooked a bloodstained carpet, a bloody mop, and a chandelier—not yet installed, and sitting in an open box—that was smeared with blood and fingerprints. In hindsight, District Attorney

Dossett acknowledged, the initial search should have been conducted by trained investigators, not lawyers.

With the discovery of the body in the bathroom, the missing car took on new significance, and the police reported it stolen. As it turned out, the man who had stolen it was already in custody, some three hundred miles to the south, in Columbus, Georgia.

On Monday, January 28—the day Barnett failed to return to work—an acquaintance of his was visiting Columbus and spotted what appeared to be Barnett's car, a blue 1982 Oldsmobile with a Knox County tag. Suspecting something was amiss, the sharp-eyed Knoxvillian called the Columbus Crimestoppers hotline, giving the car's description and location. The car hadn't yet been reported stolen, but the Columbus police acted on the tip anyway, and pulled it over. The man behind the wheel—Jeffery Charles Middlebrook—quickly explained that his friend Ted had loaned him the car. But when the police checked his records, they found that he was wanted on burglary charges, so they took him in and detained him. On December 31, when the car was reported stolen, the Columbus police called Knoxville, and two homicide investigators, Tom Stiles and Gary Moyers, drove down to talk to Middlebrook.

Middlebrook admitted to the detectives that he had killed Barnett. He gave two explanations why. One statement indicated that robbery was the motive; in the days after the murder, he said, he sold and gave away clothes and furnishings from Barnett's house. (One Knoxville man who went to KPD for an interview several days after the body was discovered actually arrived at KPD wearing a shirt that had belonged to Barnett.) Middlebrook also acknowledged stabbing Barnett repeatedly over several days, to see if he was still alive. Art had found evidence of partying in the house, and Middlebrook confirmed that he had not been the only one in the house with Barnett. Two women and two other men were with him, he said, though he gave conflicting accounts of whether anyone else was involved in the slaying. In one statement, he claimed another man helped him kill Barnett; in another, he claimed that no one else even knew Barnett was in the house.

Middlebrook's other statement to Stiles and Moyers hinted at a darker motive for the killing, and a darker side to the prosecutor's private life. Barnett made sexual advances, Middlebrook claimed, and wouldn't take no for an answer. A struggle ensued, and Middlebrook claimed Barnett was getting the better of him. That, he said, is when he stabbed him for the first time.

The story rang at least partially true. Ted Barnett wasn't openly gay—for an assistant district attorney in 1987, coming out of the closet could have been a costly career move—but friends and even some of his colleagues suspected he was gay. What's more, during Art's work at the crime scene, he found a letter Middlebrook had written to Barnett months before the killing—a letter Middlebrook sent from Brushy Mountain State Penitentiary, where he was incarcerated at the time. "Dear Teddy Bear," the letter began. Such intimate correspondence painted a startling picture of a prosecutor's relationship with a convicted criminal. The part of Middlebrook's statement that rang less true, in light of his affectionate nickname for Barnett, was the implication that he was surprised or offended by sexual advances from the attorney.

Still, despite some inconsistencies in his statements, the fact remained that Middlebrook had confessed to murdering Barnett. The case, it would seem, was open and shut.

But just as there's many a slip between the cup and the lip, there's often a snag between the confession and the conviction. As months passed, that would prove true yet again in this case.

Normally, a homicide in Knox County would be prosecuted by the district attorney and his staff. But not only was a member of the D.A.'s staff the murder victim in this case, the D.A. himself—along with several members of his staff—had gotten directly involved in the case when they searched Barnett's house and eventually found his body. As a result, they had not one but two conflicts of interest: first, their potential bias against the admitted killer of their colleague, and second, their unusual role as witnesses. D.A. Ed Dossett had no choice but to recuse himself, along with his entire staff, from the case.

In such circumstances, the Tennessee Association of Attorneys General—the network of all the D.A.'s in the state—assigns a prosecutor from another judicial district to take the case. This time, the association's executive director called on Paul Phillips, the district attorney from the Eighth Judicial District, which encompasses several rural counties to the north and northwest of Knoxville.

To hear Paul talk, you might think he's a simple country boy, but his hillbilly drawl is deceptive. By the late 1980s, Paul was one of the acknowledged stars among the state's prosecutors. Smart as a whip, he'd earned his law degree at Vanderbilt, and had been elected to his first term as D.A. at the tender age of thirty, the minimum allowed by law. (Paul has a brother who's a federal judge in East Tennessee, so clearly something in the Phillips genes or household tended to produce legal eagles.)

I had worked with Paul on several cases, and his quick grasp of forensic anthropology had impressed me strongly. One of the first cases he consulted me about was the murder of a middle-aged woman in Scott County, a mountainous county in the heart of the coal-rich Cumberlands. According to family and friends, the woman was undergoing something of a midlife crisis, and had taken to hiking for hours in the hills around Huntsville, the county seat. (One of Huntsville's claims to fame is that it is the home of former U.S. Senator Howard Baker, who rose to prominence during the Watergate hearings, which culminated in the resignation of President Richard Nixon. Baker later served as White House chief of staff, and later still as U.S. ambassador to Japan.) The Huntsville woman had gone missing, and in the course of searching for her, sheriff's deputies searched the rubble of a freshly burned shed in the mountains—the smoke from the fire had been visible from the courthouse, Paul told me. The deputies didn't find anything in the rubble, but when the woman remained missing for a second day, they went back for a second search, and this time they found some badly burned bones. Paul's question in that case was, could I determine whether the bones were human and whether they belonged to a woman. I assured him that both were possible; in fact, I was also able to make a positive identification of the bones, which did indeed belong to the missing woman. At some point during her midlife crisis she began meeting a man—a local ne'er-do-well—for sex at the shed. Eventually, he told police, he tried to break off the affair, but she threatened to cause trouble for him (her husband, a local contractor, was also a county commissioner with considerable influence in the community). So he killed her, piled kindling around the body, and set the shed on fire. He nearly got away with it, too.

In the Ted Barnett case, Paul called me for help as the trial was drawing near. After Middlebrook confessed to the murder and returned to Knoxville, the court appointed a defense attorney, Gordon Ball, who sought to suppress the defendant's incriminating statements. If the defense attorney's strategy proved successful, Paul would be forced to build his case from scratch, essentially—a task that would be greatly complicated by the autopsy the Knox County medical examiner's office had performed on Ted Barnett's body.

The M.E. at the time was a physician named Randy Pedigo. Dr. Pedigo was a smart and capable doctor, but he was a surgeon, not a pathologist, and his usual patients were alive, if not exactly well. Barnett's body bore dozens of stab wounds; in addition, the corpse was too cold to allow an easy, temperature-based estimate of TSD. When a body is still warm, it's possible to estimate TSD by taking the corpse's internal temperature with a rectal thermometer. As a general rule of thumb, a body's temperature drops by roughly 1.5 degrees Fahrenheit every hour after death, so a corpse whose temperature measures 83 degrees, for example, has likely been dead somewhere around ten hours. If the body's in a warm environment, obviously, it won't cool that quickly; Ted Barnett's body, on the other hand, had cooled to the temperature of the bathroom where it was found—a temperature that one of the investigators likened to a meat locker—and there was no way to calculate how long its temperature had matched the bathroom's ambient temperature. What's more, the cold had also slowed the chemical processes of decomposition, so the body had not yet progressed to the "bloat" stage of decomposition, when internal gases cause the abdomen to swell. Outdoors in midsummer, the combination of heat, humidity, and insect activity can reduce a body to bare bone in less than two weeks in East Tennessee. In an uninsulated room the temperature of a meat locker, though, that same body could last for months without dramatic decay.

In his initial examination of Ted Barnett's body at the death scene, Dr. Pedigo had speculated that death could have occurred weeks prior to the body's discovery. But he enlisted a pathologist at UT Medical Center, Dr. John Evans, to perform the autopsy. I don't know why Pedigo didn't ask me to weigh in as well, since by that time I'd been researching human decomposition for seven years at the Body Farm, which was located only a few hundred yards from the hospital—and which Pedigo had visited several times. Perhaps he didn't consult me because of the sensitive nature of the case. In any event, in his autopsy report, Dr. Evans estimated the TSD at five to eight days.

The problem with Dr. Evans' estimate was that Jeffery Middlebrook had witnesses who could testify that he was in Columbus, Georgia, for part of that period. And he had an ironclad alibi beginning December 28, when he was stopped by the police at the wheel of Ted Barnett's blue Oldsmobile. A good defense attorney—and Gordon Ball was a very good defense attorney—might be able to use the witness testimony and even Dr. Evans' TSD estimate to create reasonable doubt in the minds of a jury.

Paul Phillips didn't tell me the prosecutorial problems the autopsy's TSD estimate created for him; he simply mentioned the discrepancy between Dr. Pedigo's assessment and Dr. Evans', and then he asked me two questions: Could I comment on the difficulty in estimating the length of time since death? and—drawing on my research on human decomposition—could I make a scientifically reliable estimate of when Barnett was killed?

By this point, nearly eighteen months had passed since Ted Barnett's bound, wound-riddled body had been found in that chilly bathroom. All I would have to go by would be the crime scene photos, weather data, and the recollections of investigators who had worked the crime scene. It would be a challenge, but I assured Paul I would do my best.

Paul's criminal investigator, Normal Acres, rounded up everything that might provide me with data. The files included thirty-one color photographs taken at the crime scene the day the body was discovered; fifteen color Polaroid photos of the house, yard, and openings under the house, taken later by Acres; a copy of Dr. Evans' autopsy report; the KPD's drawing of the bathroom layout (including the body's position); notes from an interview with Detective Gary Moyers, one of the investigators who had worked the death scene alongside Art Bohanan; and a National Weather Service report, "Local Climatological Data Monthly Summary," for December 1987. In addition, Dr. Evans was kind enough to loan me his complete file, which included additional color photos he took at the time of the autopsy.

The National Weather Service data was almost enough to make me shiver. During the period between December 16 and December 31, 1987, there were five days in which the average temperature was in the thirties, six days in which the average temperature was in the forties, and four days in which the average temperature was in the fifties. Besides corroborating Detective Moyers' impression that the bathroom was as cold as a meat locker, the temperature data told me that the body would have decayed very little during that sixteen-day interval. What's more, besides slowing the chemistry of decomposition—the assaults of bacteria and the body's own cellular self-destruction, called "autolysis"—the cold had shut down the other key contributor to the corruption of the flesh: insects. Maggots, which are the wormlike larvae of blowflies, normally play a huge role in human decay, consuming (in one study at the Body Farm) up to forty pounds of soft tissue per day.

But maggot feeding frenzies of that magnitude occur only when the weather's warm and blowflies have unrestricted access to a body. Ted Barnett's roughly finished bathroom had enough gaps in the floorboards for blowflies to get in, but at those temperatures the insects wouldn't have tried: blowflies are dormant if the temperature is below 52 degrees. In short, judging by the state of decomposition depicted in the crime scene and autopsy photos, Ted Barnett might just as well have been in the cooler at the morgue.

But the photos were far from useless; they contained dramatic visual evidence I realized could be compared to research corpses at the Body Farm. The crime scene pictures showed mold colonies growing on the right side of Barnett's face and on his neck. In addition, Dr. Evans' autopsy photos revealed additional mold on the chest and the feet. The mold colonies on the face were small to midsized, with the largest being roughly the size of a quarter; those on the neck were larger.

As I studied the crime scene photos, I saw other mold colonies in the bathroom as well. There were large patches of mold under the sink (Barnett's head lay near the base of the toilet and partway under the sink) and in the shower (where his feet were). Detective Moyers had reported, "There was water running in the toilet, water dripping in the sink and there was a slight drip of water in the shower stall." Clearly Barnett's body was in a prime environment for mold growth—except for one thing. Mold growth is like decomposition: it happens faster in warm environments, slower in cool environments. So the fact that sizable mold colonies had formed on Barnett's body in that cold bathroom meant that quite a bit of time had elapsed since his death—considerably more, I felt sure, than the five to eight days Dr. Evans had estimated.

For a more objective perspective, I consulted my files of color slides of decaying corpses at the Body Farm. I looked at photos of our first three research bodies, which decayed in the spring, summer, and fall of 1981: the bodies numbered 1-81, 2-81, and 3-81. For consistency, I looked at photos taken fifteen days after the bodies were placed in the research enclosure. Body 1-81 showed no indication of mold; body 2-81 showed extensive mold; and body 3-81 showed a medium-sized colony on the right wrist.

When I stepped back and weighed all the data that seemed significant—the lack of soft-tissue decay, the cool temperatures in the bathroom, the dampness of the environment, and the sizable mold colonies on the face, neck, and chest—I felt confident that Barnett had been dead at least ten days before his body was found, and probably closer to fifteen. In other words, he likely died somewhere around December 18. In view of the fact that he survived for several days after he was first stabbed and tied up, his fatal ordeal probably began very early in his two-week Christmas vacation, and certainly well before Christmas. Sadly, Ted Barnett was home for the holidays . . . and dead for the holidays.

Jeffery Middlebrook's trial was scheduled to begin on July 17, 1989. Thanks to the solid crime scene and investigative work by Art Bohanan, Gary Moyers, Tom Stiles, Norman Acres, and others, Paul Phillips had built a strong case against Middlebrook. Even apart from Middlebrook's admission of guilt—which his attorney was still hoping to suppress—there was no doubt that he had been drinking in Barnett's house, no doubt that he had sold, stolen, or given away many of Barnett's possessions, and no doubt that he had handled the bloody chandelier that had struck Barnett in the head sometime during the assaults that had gradually killed him. And judging from the weather data and the Body Farm's studies of decomposition and mold formation, there was no doubt that the murder had occurred well before Middlebrook had a police-certified alibi down in Georgia.

Faced with the mountain of evidence and the possibility of a death sentence, on the eve of his trial Middlebrook pled guilty to second-degree murder and armed robbery. He received a sixty-year sentence: thirty-five years for the murder charge and twenty-five years for armed robbery, the maximum allowed. According to the Tennessee Felony Offender Information Lookup, a web site run by the state's Department of Correction, he will be eligible for parole in 2041.

7

A TEXAS SCORCHER: FINDING A "FINGERPRINT" IN A FRONTAL SINUS

The guy at the junkyard in Corsicana grinned and shook his head when I told him why I was there. "There ain't nothing in that car," he said. "I've looked."

"Well, if you don't mind, I'd like to take a look anyhow," I said. "Long as I'm here." He didn't mind; in fact, he seemed to find the prospect amusing. He opened the gate and I drove inside in the car I'd rented at the Dallas airport.

I could almost hear what the guy was thinking: *People from Tennessee—hell, even the college professors—are idiots.* What I was thinking was, *We'll see about that.*

Of all the hundreds of forensic cases I've worked over the years, few have been as challenging as a handful in which I was asked to identify someone on the basis of small fragments of incinerated bone.

Killers sometimes think they can destroy a body completely by burning it, but the truth is, that's extremely difficult to do. In one memorable case, when my graduate students and I searched the rubble of a house that had burned down with such fierce heat that even the copper wiring melted, we still found an identifiable skeleton . . . with a melted bullet lying alongside the spine. We could even tell that the body had been blown apart with a stick of dynamite before the house was torched. I'm telling you, a body is hard to burn. If you decide to murder somebody, don't think that you can cover your tracks with fire.

Even a cremation furnace—a high-temperature, gas-fired apparatus designed specifically and *solely* to incinerate human bodies—doesn't completely destroy a body. True, by subjecting corpses to temperatures of 1600 to 1800 degrees Fahrenheit, cremation furnaces do destroy all the organic (carbon-containing) chemicals in the body, including DNA; however, when the gas is turned off, the furnace is cooled, and the door is opened, what remains is a recognizable human skeleton. It's fragile and brittle, because it's been "calcined"—reduced to nothing but calcium and other minerals, without the collagen matrix that binds the minerals together so strongly. But it takes another step to reduce those calcined bones to the granular, ashy powder funeral directors call "cremains," and that step involves feeding the bones through a pulverizer, a machine with a powerful, blender-like blade that chops or grinds the brittle bones to fragments.

In another case, years before I found myself on the receiving end of that patronizing grin in Corsicana, I was asked to examine some burned bones from a house that had been destroyed by fire. The bones had been identified by a forensic pathologist as dog bones but the homeowner was nowhere to be found, so the sheriff wanted a second opinion. I could tell immediately that the bones were human, although there was nothing in what I received—a fragmentary pelvis, the lower lumbar vertebrae, and parts of both femora (femurs)—that could be used as the basis for a positive identification. I asked the sheriff if I could go to the burned house where these bones had been recovered, in hopes that I could find the fragmented skull and make an identification. He agreed to take me there himself—it was in a rural part of Middle Tennessee, and I was glad he knew the way. Not only did we find enough pieces of skull to make a positive identification, we also found a .32 caliber pistol (which the sheriff said the dead man "always carried"), a handful of brass rivets from blue jeans (which he "always wore"), and three empty cans of charcoal lighter fluid. We did *not*, on the other hand, find any coat hangers in the closets, any dishes or silverware in the kitchen, or any remnants of burned furniture. So five weeks after the fire, at a scene that had already been investigated, there remained enough undiscovered evidence to identify a dead man, and to show that his house had been emptied and then torched—probably by the man himself, in a fatally botched insurance scam.

All this is by way of background, to explain why I offered some hope to a State Farm insurance investigator, Larry Patterson, who called me in June of 1989 to ask if I might be able to tell whether or not the incinerated remains in a burned car were those of a man who held a $250,000 life insurance policy. Both the car and the policy belonged to a man named Owen Rutherford; the burned vehicle was found in a distant corner of a 25-acre field on Rutherford's property outside Corsicana; and the man had not been seen since one day in February, when he announced that he planned to do some drinking to forget the financial worries that were plaguing him at the time. On the face of it, it appeared plausible that the burned remains were indeed Owen Rutherford's. However, insurance companies don't like to pay out big life insurance claims on the basis of mere plausibility, and they've got good reason to be cautious: if Owen Rutherford had faked his death in a car fire, he wouldn't be the first person to try that trick, nor would he be the last, even in my own limited experience.

When the charred shell of Rutherford's car was searched that February, a badly burned torso was found inside, along with a few other incinerated parts. These were sent to the Southwestern Institute of Forensic Sciences—the Dallas County crime lab—for examination. A forensic pathologist there, Dr. Violette Hnilica, autopsied the remains, X-rayed the charred torso, and attempted to match the images to X-rays taken of Owen Rutherford during his life. She did not succeed in finding a match, so she was unable to confirm that the remains were Rutherford's.

That's when State Farm called to ask if I could help. As a first step, I asked to see both sets of X-rays, antemortem and postmortem. Through a judge in Corsicana, the insurance company arranged to have the films sent to me from Dallas. They arrived on July 31; I signed an evidence receipt to confirm I had them, and mailed it back to the crime lab, keeping a yellow copy for my own files.

When I compared the sets of X-rays, I could see why Dr. Hnilica had been unable to make a positive identification: the best antemortem X-rays were of

Owen Rutherford's head, and judging from the autopsy report, no head was recovered from the burned car—only a few small skull fragments. The envelope also included an antemortem chest X-ray, but the orientation of the chest in that one was different from the orientation of the postmortem chest X-ray, so it was impossible to match the two. Just to be sure, I asked one of the senior radiologists at UT Medical Center, Dr. Kenneth Rhule, to compare the sets of images as well. Ken couldn't find anything to compare directly, either. The X-rays didn't rule out the possibility that this was Owen Rutherford, mind you; they just didn't offer any basis for either a yes or a no.

When I delivered the bad news to Larry Patterson on August 1, I cushioned the blow with a glimmer of hope: the burned vehicle might contain some additional skeletal material; with luck, maybe even a piece that would allow a positive identification. "The best means of identification is at the death scene," I said, repeating a mantra I've chanted countless times to police and forensic technicians. Patterson asked me to talk to his boss, a claims administrator named Chris Johnson, so I did. "If I could search that vehicle myself, I might be able to find something," I told her. Finally she agreed to fly me and Murray Marks, one of the Ph.D. students on my forensic response team, to Texas to see what we could find.

Corsicana was a small city (about 25,000 people) sixty miles southeast of Dallas. One of its claims to fame was the discovery, back in 1894, of the first significant quantities of petroleum west of the Mississippi. The discovery was purely accidental—the surprised well-drillers were actually looking for water—but it launched the Texas oil boom, nearly doubled Corsicana's population, and created a thirst not just for oil but for whiskey, which thirty-five saloons sought to slake. A later claim to fame was Corsicana's contribution to U.S. airpower during World War II: between 1941 and 1944, flight instructors at Corsicana Field trained more than 8,000 Air Force pilots, including P-38 fighter pilot Tommy McGuire, who shot down thirty-eight Japanese planes, the second-highest number of kills by any American ace, ever.

Corsicana is the seat of Navarro County, and one of its grandest buildings is the courthouse, built in 1905 of red granite and gray brick, monumental in the way Texas courthouses tend to be. The first stop Murray and I made in Corsicana was across the street from the courthouse, where we briefed Justice of the Peace Badie Stewart on how we would search the burned vehicle, and what we hoped we might find. Owen Rutherford had reportedly recently been through a "sticky divorce," according to Judge Stewart; he was also broke, with creditors closing in around him. But Rutherford was also a bit of an outlaw, Judge Stewart said, then surprised me by adding, "It won't surprise me if it's not him."

It was 11 A.M. and already hot as hell—it was east-central Texas in August, after all—when we arrived at Haynie Wrecker Service and met the dubious Mr. Haynie. The burned vehicle was a two-door Pontiac 2000, which had been towed to the fenced-in area behind the junkyard's office. The car was, as the saying goes, a mere shell of its former self. The doors and trunk lid, and even the roof, had been cut off—removed with a torch at the scene—to allow the torso and other remains to be retrieved. Haynie also told me that he'd known Owen Rutherford, that Rutherford had a gap between his two upper front teeth, and that the remains removed from the car had a similar gap. This revelation came as quite a surprise to me, because the autopsy report hadn't mentioned any teeth, and I had received no postmortem X-rays of teeth.

The wreckage was surrounded, and practically hidden, by waist-high grass and weeds. After doing a quick check for snakes, Murray and I began combing the debris that had collected in the vehicle's floor pan (an excellent, though unintentional, evidence-retention feature of vehicles). It wasn't long before we hit pay dirt—or bone, rather. In the area of the driver's seat, we found bone fragments from the hands and feet, as well as the distal end of the left fibula. As Murray continued to excavate, I began sifting material through a wire screen I'd brought. The material—a mixture of dirt, ash, embers, rusty metal, and bone fragments—was wet and crumbly, so it was a bit like teasing cookie dough through the screen.

Just behind the right front seat, Murray started finding more bones. These were phalanges, or finger bones, from the right hand, as if Owen Rutherford, or whoever it was, had died and burned with his arm draped over the seat back. In the rear behind the driver's seat—actually in the backseat, where the seat had fallen through to the frame—Murray began finding what I'd particularly hoped we might find in the vehicle: pieces of the cranium. The biggest and easiest to identify was the right zygomatic arch, or cheekbone; the other pieces, about fifty in all, were fragments that ranged from small (roughly the size of a quarter) to tiny. The skull consists of three layers: a hard outer layer, a spongy middle layer (called "diploë"), and another hard inner layer. In a particularly hot fire, the air spaces and moisture in the diploë can create tiny pockets of steam and pressure, and these can cause the layers to separate. That's what happened in this car fire: of all the dozens of bone fragments from the skull, none were intact through all three layers; instead, all had fractured in the diploë.

At midafternoon, as Murray and I mopped our brows and prepared to wrap up our excavation, we suddenly hit the mother lode. The trunk of the car contained enough burned human material to fill a five-gallon bucket—numerous pieces of ribs, long bones, fingers and toes, though not, alas, any teeth or pieces of skull. This treasure trove of bones and fragments was wrapped in a sheet labeled "Corsicana Regional Hospital," which is where Dr. Hnilica, the pathologist, had conducted the autopsy. I didn't understand why skeletal material from an active death investigation had been dumped back in the trunk of a burned car in a junkyard, but then, as Mr. Haynie had intimated, I was just a moron from Tennessee. How could I possibly grasp the sophisticated forensic ways of a big-league state like Texas?

By the time Murray and I left Corsicana at three-thirty that afternoon, we'd recovered and boxed up nearly 300 pieces of bone from the car, including 18 rib fragments, 4 skull fragments, 78 bones or fragments from the hands and feet, 40 pieces of long bones, 1 left patella (kneecap), and 97 fragments too small to identify. But would any one of those pieces tell me, with certainty, whether or not the body, like the car, belonged to Owen Rutherford? I pondered this on the plane ride back from Dallas that night. The round-trip, plus the excavation, had made for a long damn day, but I was glad to be out of Texas and back in Tennessee.

Back in the bone lab beneath Neyland Stadium, Murray and I began sorting through the material we'd brought back from Corsicana. To make sure we had pieces of only one person, we laid out the fragments—the two hundred we could identify—in anatomical order, looking for any duplicates. There were none, so I was confident that the fire had claimed only one victim. The pieces were shattered and distorted, as is normally the case from a hot fire. Exposure to extremely high temperature causes the moisture in fleshed bone to turn to steam; the steam leaves the bone so fast that the bone itself shrinks, twists, and splinters. Indeed, the pieces of long bone—fragments from the femur, tibia, fibula, radius, ulna, humerus, and bones from the hands and feet—had twisted, split, and even shattered from the pressure of the steam in the marrow cavities. Other bones, including pieces from the cranium, vertebrae, ribs, pelvic girdle, and the base of the foot, had come apart at the seams, where an intermediate layer of spongier bone was sandwiched between two harder layers. Obviously this extreme damage from the fire complicated the effort to make a positive identification. It also pointed to an extremely hot fire—and made me wonder if accelerants were used, and if we were looking at a murder rather than an accidental death or suicide.

The bones were obviously human. From the large size of the patella (kneecap), it seemed likely that the sex was male. A bit of arthritic lipping on the bones of the foot suggested that he was middle-aged, as Rutherford had been. Nothing so far contradicted the possibility, or even the likelihood, that this was Rutherford.

Unfortunately, State Farm and Judge Stewart needed more than possibility or likelihood; they needed certainty, a positive identification, and that was proving positively elusive. We'd found no teeth or even fragments of teeth in the car—a surprising omission, considering what the junkyard's owner had told us about a gap in the burned body's teeth. With what we had, our only hope of determining whether this was Owen Rutherford lay in the few fragments we'd found from the cranium and face. All we needed to find was one distinctive, identifiable feature or shape we could match to an antemortem X-ray of Rutherford. Try as we might, however, we couldn't find it: not a single diagnostic structure or form that would allow a positive ID.

On August 18, nine days after our whirlwind trip to Corsicana, Murray and I signed a report detailing our inconclusive findings. We sent it to Larry Patterson and Chris Woodson at State Farm and to Navarro County Justice of the Peace Badie Stewart. There might be one final possibility, we noted at the end of the report. Out in Corsicana, we'd learned that after the original autopsy, parts of the body (including the torso) had been sent to Chicago for burial, alongside Owen Rutherford's mother. "It seems to us," we wrote, "that the only means of making a positive identification, either from dental comparisons or a change in orientation of the torso . . . were buried with the body. The buried body would need to be exhumed and careful comparisons made with before-death X-rays."

On August 23, I spoke with Dr. Hnilica, who delivered yet another piece of surprising news, this one welcome. The Dallas crime lab had retained some of the skeletal material the forensic techs had recovered from the burned car, including half of the mandible and the teeth associated with it. The lab's senior field agent, Rick George, was in charge of the skeletal material, so I hastened to contact him to request whatever he had.

On October 6, I got a carton from Dallas containing two plastic bags with fragmented and incinerated human skeletal remains. Yet another unexpected development awaited me: the bags contained 101 identifiable fragments of skeletal material, including 66 pieces from the cranial vault, 17 from the roots of various teeth, and 18 from the mandible, as well as numerous crumbs of bone too small to identify.

What excited me most were eight small jagged fragments from the frontal bone, or forehead. Murray and I managed to reassemble these, like tiny pieces of a miniature jigsaw puzzle. When we were finished, we had a piece of frontal bone that spanned much of the forehead, including the prominent brow ridge above the eyes. That region of the frontal bone also contains the frontal sinus, one of four sets of air-filled cavities in the bones of the face and skull. It's not entirely clear why we have sinuses, aside from their role in providing a huge revenue stream for pharmaceutical companies and over-the-counter drug makers. Some scientists think sinuses evolved mainly to lessen the weight of the front of the skull; others, to protect the eyes and brain against physical blows or rapid temperature changes; still others, to humidify air as we breathe in. An even more out-of-the-box theory holds that their role is to give the human voice more resonance.

Forensically, what matters isn't so much *why* we have frontal sinuses; what matters is *how* we have them: uniquely. Just as the loops and whorls and arches of fingerprints vary from person to person, so do the cavities and edges of frontal sinuses. In Owen Rutherford's antemortem X-rays, the upper edge of his frontal sinus appeared scalloped, rather like the ends of leaves from a ginkgo tree, with a wavy vertical line—the sinus's septum—meandering up the center of his brow ridge to divide the right side of the frontal sinus from the left. When we X-rayed the portion of frontal bone we'd pieced together from the eight burned fragments, I felt a thrill of excitement. The wavy line of the septum tracked Owen Rutherford's exactly; so did three of the scallops marking the edges of the sinus cavity.

That wasn't all. The pieces of mandible that arrived in the package contained two impacted third molars—wisdom teeth—which were visible to the naked eye, since the soft tissue around them had been burned away. Owen Rutherford's antemortem X-rays clearly showed these same two impacted molars, in exactly the same orientation. Judge Stewart's cryptic comment—"It won't surprise me if it isn't him"—had hinted that there might be more to this case than met the eye. And maybe there was, but I hadn't been asked to figure out whether this man had committed suicide, had passed out and dropped a lit cigarette onto the car's upholstery, or had been murdered. All the insurance company and the judge wanted to know was, "Was it Owen Rutherford who burned up in that car?"

On October 11, I sent a second forensic report to State Farm and Judge Stewart. Yes, I said, it was Owen Rutherford. Positively.

I wish I had sent a copy to the guy at the junkyard, too.

8

NO BUTTS ABOUT IT: CONVICTING A KILLER BY MATCHING BITE MARKS

By 1991, I had spent thirty-five years studying bones. I had learned how to interpret blunt-force trauma—for example, the way skull fractures radiate from the spot where a hammer or an iron pipe has slammed into a victim's head. At a glance, I could recognize the conical beveling that made a bullet's entry wound wider on the inside of the cranium than on the outside . . . and that made the exit wound (which was nearly always larger than the entry wound) bigger on the outside of the skull than on the inside. I could spot—or feel—tiny nicks in ribs marking where a thin blade had slipped through the protective cage of bone to pierce the heart or lungs.

I felt at home with bones. But when FBI Special Agent Don Provonsha asked me to examine the evidence in a high-profile Knoxville murder case in 1991, he didn't bring me bones, or flesh, or even insects. He brought me the butts of two small cigars, which the FBI had labeled Q65 and Q124–133. The butts of these ordinary, inexpensive cigars, Hav-A-Tampa Jewels, could make or break the government's case.

One of the butts, Q65, had been found near the spot where a Tennessee Department of Public Safety officer had been gunned down at point-blank range. The murder was a top priority for the Bureau, because the victim, Michael Rector, was working for the FBI as an undercover informant. Rector was the linchpin of an eighteen-month racketeering investigation that was bearing down on one of the biggest automobile "chop shops" in the Southeast—a chop shop owned by Rector's own brother-in-law, Roy Lee Clark.

West Wood Enterprises—a junkyard and used auto-parts business—occupied several acres of hillside and hollow just outside the Knoxville city limits. The area was known as Rocky Hill because it was elevated and, well, *rocky.* Although Knoxville's suburbs were creeping in that direction, Rocky Hill retained a fairly rural character, with small frame houses and the occasional mobile home rubbing shoulders with a few larger, grander homes set on multi-acre tracts. West Wood Enterprises was situated off the lower end of Derris Drive, a narrow ribbon of unstriped asphalt that snaked up a wooded ravine between Wright's Ferry Road and Duncan Road. In addition to hundreds of cars and trucks in various stages of disassembly, the West Wood compound held several motley buildings: a two-story brick ranch house (whose second-floor windows had been bricked up); another two-story brick house, this one L-shaped (with a high-roofed garage tacked awkwardly onto one end); a long series of adjoining sheds; and, farther up the hillside (and farther from prying eyes), a barn-sized garage and workshop. The property was surrounded by woods, and the only way in or out was a gravel driveway protected by a chain-link gate.

Roy Clark and his wife Carol lived up on Duncan Road, about a mile away and considerably higher both topographically and architecturally. The Clarks occupied a substantial house set on the brow of a hill. Their driveway was gated, too, not with chain link, but with a pair of wrought-iron gates, each monogrammed with a foot-high C.

Carol Clark's brother, Michael Rector, had worked briefly for the Knox County Sheriff's Office in the 1980s. After losing that job, he'd gone to work for Roy—stealing and dismantling cars—and had lived in one of the oddly customized buildings on the West

Wood property. Then he'd gotten a job with the Public Safety Department's Motor Vehicle Enforcement Division, which inspected commercial trucks for excessive payloads and other safety violations. Rector was assigned to the truck scales on Interstate 40—not the most glamorous job in law enforcement, but it was honest work, and it paid the bills. Rector and his wife Sally lived in a middle-class subdivision in West Knoxville. Their home, on a corner lot at the intersection of Penwood Drive and Sedgewick Road, was a tidy two-story split-foyer rancher. A two-car garage occupied one end of the basement; across the back, a large, elevated deck ran the length of the house, looking out on a backyard dotted with maple trees and white pines. It was a sizable step up from Rector's former quarters at the junkyard.

In April 1988, Rector had contacted the FBI and offered to provide information about illegal activities by his brother-in-law. It was exactly the sort of inside help the agency needed: Clark had been under scrutiny by local and state authorities for years, but building a case against him had proved difficult—perhaps partly because several law enforcement officers in East Tennessee had reportedly purchased vehicles from Clark at bargain-basement prices. But by the late 1980s, his sales of vehicles and parts had expanded beyond Tennessee into neighboring states. If he was trafficking in stolen property across state lines—as Rector said he was—Clark was vulnerable to federal racketeering charges.

For a year and a half, Rector held secret meetings and telephone conversations with agents detailing how Clark stole cars, disassembled them in his chop shop, and sold the parts. Sometimes Clark reversed the process, too, Rector said, assembling vehicles out of stolen parts; occasionally he simply resold the stolen vehicles, after destroying or altering their serial numbers.

By December of 1989, the FBI had laid the foundation for a massive case against Clark. The stakes were high, and not just because Clark was running a huge chop shop. This would be the Justice Department's first use of the RICO racketeering statutes to prosecute a property-theft case. To cement the case, Provonsha asked Rector to wear a hidden tape recorder—a wire—and record incriminating conversations with Clark. Rector agreed. Over the next three months, he taped Clark repeatedly—sometimes at West Wood, sometimes in Roy's car or Rector's truck, as the two men rode around scouting for cars for Clark to steal. On December 28, Rector recorded Clark in the act of stealing a vehicle, which he then took to West Wood. The information on the recordings could potentially put Clark away for years. But the FBI was taking no chances; the agents wanted an ironclad case.

In one of the taped conversations, Clark asked Rector to help him find a Chevy Blazer. Not long after that, Rector did. On March 2, 1990, he led Clark to a West Knoxville motel where a black Blazer was parked. Later that evening, Clark drove into the motel's parking lot in his Cadillac Coupe de Ville, parked beside the Blazer, and got out. He took out a flat metal tool called a "Slim Jim"—used by tow truck drivers, police officers, and car thieves to open locked doors—and slid it down between the driver's window and the doorframe. He wiggled the tool back and forth and then gave a slight upward tug, popping the Blazer's lock. Then he opened the door, peered inside, and closed it again, without locking it. He got back into his Cadillac, drove around to the other side of the motel, and parked. A few moments later he walked to the Blazer, got in, and began fiddling with the steering column. In less than thirty seconds the Blazer started, and Clark drove it away to West Wood, where it was dismantled with equal efficiency.

The Blazer's owner witnessed the theft from start to finish—witnessed it, photographed it, and videotaped it. The Blazer's owner was the United States government—the Federal Bureau of Investigation, to be specific. Clark had just stolen federal property, and a cluster of FBI agents in a second-floor room of the motel had watched in amazement at the swiftness and coolness with which he'd done it.

The Bureau could have swooped down on him immediately. Instead they chose to keep biding their time and building their case. Clark's automotive wish list also included a black Cadillac Eldorado. On March 23, Rector obligingly found him one, again with the help of the FBI, which had parked the car at the Club House Inn. Once more, as Clark pulled up beside the government's lure, Provonsha and other agents were watching through night-vision lenses.

When Clark pulled out his Slim Jim, Provonsha started timing. It took Clark barely ten seconds to open the locked door. As before, he bent over the car's steering column. This time, however, the car did not start swiftly. Several minutes passed, and then—to the agents' surprise—Clark emerged from the Eldorado, closed the door, and walked away, scowling. The Cadillac's antitheft device was better than the Blazer's. Roy Lee Clark had met his vehicular match.

Clark had met his investigative match as well. On April 1, a caravan of unmarked FBI and TBI cars, highway patrol cruisers, Knox County sheriff's cruisers, and Knoxville Police Department cruisers turned up the narrow asphalt of Derris Lane, ascended the gravel driveway of West Wood Enterprises, and descended on

Roy Lee Clark like the wrath of God, or at least the wrath of the criminal justice system.

Accompanying the agents was Mike Rector—a fact whose significance might not have been clear to Clark, since numerous other state officers had been assigned to the bust as well. Over the next five days, guided by Rector, law enforcement agents swarmed over the hillside junkyard like flies around a corpse. Rector pointed out stolen vehicles as well as parts that had come from vehicles dismantled in the chop shop. Chaperoned by a state trooper, Clark monitored their progress from a distance, pausing occasionally to urinate in the weeds and speculate about who had sold him out. By the end of the search, the FBI had hauled off five tractor-trailer-loads of evidence, valued at $700,000: engine blocks, fenders, doors, bumpers, wheels, and transmissions, as well as boxes of documents detailing West Wood's transactions with customers in Tennessee, Georgia, Kentucky, Alabama, and other states.

At the end of the five days, Mike Rector walked up to his brother-in-law—the man who had turned him, for a time, into a car thief—and said, "It's me, Roy. I'm the one who's been working with them. I wanted you to hear it from me, not them." Clark looked stunned.

The federal prosecutor, Assistant U.S. Attorney Russ Dedrick, gave Clark a deadline: negotiate a plea bargain by June 1, or take his chances with a jury. With the audiotapes, videotapes, photographs, seized documents, and Rector's damning testimony, Clark's chances in a jury trial didn't look good.

On the night of May 30, Roy and Carol Clark—driving separate cars—rendezvoused at Naples, an Italian restaurant a few miles from their home. After dinner, they drove their cars to the home of some longtime friends, Richard and Carol Maples. Carol Clark drove up the long gravel driveway and parked near the house; Roy parked out on Bluegrass Road, roughly a hundred yards away. No one was home at the Maples house, but the Clarks let themselves inside anyway. Late in the evening, Richard and Carol Maples returned home to find the Clarks waiting for them. Over the next several hours, as "the Carols" visited in the house, Roy drifted in and out of the house and the garage, where Richard and a friend began working on a car, a vintage street rod.

Meanwhile, Mike Rector was finishing up his shift at the I-40 truck scales. At 11 P.M., Rector clocked out and climbed into his pickup truck, but as he started down the ramp to the interstate, he realized that one of his tires was flat. To make matters worse, the spare was at home in the garage. Rector reparked the truck, went inside the scale house, and got permission to borrow one of the state's cruisers to go home and retrieve a spare. His home was half a mile from the house of Richard and Carol Maples—the house where Roy Lee Clark was coming and going, indoors and out.

Two hours later, having returned the cruiser and changed the tire, Rector turned onto his street in his truck. He stopped at the curb to collect the mail from the mailbox. Then he backed into his driveway, parked, and got out. He started up the sidewalk leading from the driveway to the front door, carrying the mail and his gun belt, a Smith & Wesson semiautomatic pistol snapped securely into the thumb-break holster.

The assailant's first shot caught Rector under the chin, perhaps as Rector was looking his killer in the eye, maybe even talking to him. Four more bullets tore into him, including two in the back. Rector fell facedown in the grass beside the sidewalk, his arms tucked beneath him, his right foot behind his left ankle. The day's mail landed on the sidewalk, still bundled neatly together, and the officer's unbuckled gun belt sprawled beside the mail. Two red circles on Rector's shirt marked the gunshot wounds in his back. He lay there unseen, in his green uniform on the green grass. Sally Rector was worried because Mike was long overdue; he hadn't returned a page, and his coworkers said he'd left hours before. Shortly before dawn, Sally walked outside and found her husband's body.

Later that morning, Roy Lee Clark was arrested—not for murder, but simply for auto theft. He was denied bail, though, an unusual precaution for a car thief. Not until October 1991, nearly seventeen months after Rector was killed, did a federal grand jury indict Clark for the murder.

One reason for the delay in the indictment might have been the virtual lack of any evidence tying Clark directly to the crime: no gun, smoking or otherwise; no ballistic matches between the fatal bullets and any bullets Clark was known to have fired; no relevant fingerprints or DNA evidence; no telltale bloodstains on Clark's clothing or vehicles; no eyewitnesses. The government would need to build a strong circumstantial case against Clark. That case would hinge on motive, opportunity, and the two cigar butts brought to me by Special Agent Don Provonsha.

In 1991, DNA testing was still in its infancy; the technology was not yet sophisticated enough to detect trace amounts of genetic material in dried saliva. What made the cigar butts potentially useful as evidence was their wooden tips: tips that bore the faint impressions of teeth.

The tip labeled Q65 had been found a block away from Rector's body during a careful search of the neighborhood by FBI agents and state and local police just hours after Rector's wife found him and called 911. The unsmoked portion of the cigar appeared fresh when it was found; it was also dry, which seemed noteworthy, since a rain shower had passed through the area just two hours before the shooting.

The other cigar tip, Q124–133, had been recovered that same day from an ashtray in Roy Lee Clark's kitchen. Was it possible, Provonsha asked when he brought me the tips, to determine whether the same individual had smoked both the cigars? Was it possible, in other words, to use the cigar tips to show that Clark had been lurking near Rector's house around the time of the shooting?

One of my graduate assistants at the time seemed ideally suited to help me compare the cigar butts. Mark Gilbeau was a methodical, meticulous student with mechanical aptitude and steady hands, and if ever a project called for steady hands and meticulous work, this was it. Our first step would be to make a mold of the cigar tips, using powdered alginate—a hard, fast-setting, rubbery gum extracted from seaweed. When reconstituted with water to about the consistency of pancake batter, alginate can conform to the most minute, subtle textural details, capturing even fingerprints and skin pores. For that reason, the material was widely used in dentistry and also in Hollywood, where it was used to make molds of actors' faces, as a step in creating special-effects makeup. After making alginate molds of the cigar tips, we could then make exact replicas of the tips out of dental stone, a fast-setting, tougher version of plaster of paris. The replicas would allow us to study the tooth marks to our heart's content, without risk of damaging the original evidence.

The consistency of the alginate mixture was crucial: too runny, and it wouldn't set properly; too thick, and it wouldn't capture the subtle details that would be crucial for any meaningful comparison.

Not surprisingly, there's no custom-engineered gizmo on the market for making molds of potentially incriminating cigar tips; lacking that, we'd have to improvise. I looked around the lab and my eye happened to light on some empty 35-millimeter film canisters: small plastic cylinders about an inch in diameter and a couple of inches tall. They were the perfect size to encase the cigar tips in about a quarter inch of alginate all the way around; what's more, the cylindrical shape and slick plastic surface would make it easy to slide the molds out, once the alginate had hardened. Then we'd use a scalpel to slit the molds, opening them like a clamshell or a book, with one half capturing the contours of the cigar tips' upper surface and the other half mirroring the details of the tips' underside.

Everything was working beautifully until Mark slit open the molds and pulled the halves apart. As he did, wood fibers—weakened by the teeth that had clamped down on them—tore free from the cigar tips and clung to the hard, rubbery molds. Ironically, in trying to preserve the evidence, we had irreversibly damaged it. True, we had close-up photos of the tips, with the tooth marks sidelighted at a low angle to bring their details into dramatic relief. But a two-dimensional photograph was no substitute for a three-dimensional object. The FBI had entrusted us with the only physical evidence that might physically link their chief suspect to the crime scene, and we had inadvertently damaged it.

As Mark studied the fiber-fuzzed molds, however, he saw a ray of hope. Although fibers clung to the hard, rubbery compound, they didn't appear to be embedded in it. Mark had a pair of tiny Swiss-made jeweler's tweezers; working carefully with the tweezers, he might be able to detach the fibers without damaging the molds. Again his meticulous nature and steady hands proved invaluable; peering through a magnifying glass, he painstakingly plucked out all traces of the fibers. But were the molds undamaged? There was only one way to be sure.

Fitting the two halves of each mold back together again, Mark reinserted them into the film canisters, then drizzled a thin mixture of dental stone down into the molded-in tunnels left by the protruding portion of the cigars. The wait for the dental stone to harden seemed interminable, although it was less than an hour. Finally, when plenty of time had passed, he removed the molds from the film canisters, pried them apart, and popped out the casts of the tips. Sidelighting them just as the wooden tips had been when photographed, we inspected them under a magnifying glass, comparing them with the photos. Feature for tiny feature, texture for subtle texture, they matched the photos perfectly.

Now we could try to answer the question Provonsha had posed: Could we determine whether the tooth marks in the two tips had been made by the same person?

A casual inspection suggested they had not. The marks in Q65—the butt found in the grass near Rector's house—were much deeper than the marks in Q124–133. But as I looked closer, I noticed three distinctive notches in both sets of tips. They were deeper, to be sure, in the crime scene butt, yet the angles and relative positions and even the placement in the wooden tips were the same.

The notch in the upper surface of the tips appeared to have been made by an upper lateral incisor on the right side—the second tooth to the right of the upper jaw's midline; the notches on the underside appeared to have been made by the lower canine and first premolar. The more I thought about it, the more sense the difference in depth made: if I were a killer, waiting in ambush for my victim, I'd probably be feeling a lot of tension—tension that would probably tend to make my jaw clench tighter. The photos and the casts were consistent with this explanation: they showed considerably more damage to the wooden tip from the crime scene than the one from Roy Clark's residence.

When I told Provonsha that the bite marks matched, he was elated. It wasn't direct evidence of Clark's guilt, but it suggested, at least, that he had been near Rector's house sometime around the time of the shooting—and after the 11 P.M. rainstorm. I thought of one additional test that could be done to confirm or refute the theory that these were indeed Clark's tooth marks. "You know," I said, "I bet if we got a cast of Roy Clark's teeth, we could see if these bite marks matched his teeth." I knew a Knoxville orthodontist, Marshall Parker—he had straightened my son Jim's teeth—and I suggested Provonsha ask Dr. Parker if he'd be willing to take the impressions. Most of Parker's patients were crooked-toothed teenagers; I figured he'd consider a murder suspect an interesting change of pace.

It had been years since Marshall himself had taken dental impressions of his patients' teeth; that job fell to his hygienist, Betty Crawford. Rather than taking Roy Clark to a suburban orthodontist's office, the FBI ended up taking Betty Crawford to the courthouse. The hygienist was a bit surprised when agents led in a disheveled man in handcuffs and leg irons. She knew the man had been accused of killing someone, but she didn't know she was about to cast the bite marks of an auto-theft kingpin and notorious murder suspect; she thought perhaps the bearded, scruffy-looking man was a homeless person who'd gotten into a fight that went too far. Only on the ride back to her job did she learn from the FBI agent who the quiet, cooperative man had been.

Just to be on the safe side, the FBI also secured casts of Richard Maples' teeth.

Maples was Clark's alibi witness, after all; if Clark had done the shooting, it was possible that Maples had accompanied him, and had tossed the half-smoked cigar in the grass. It was even considered possible that Maples, rather than Clark, was the shooter. And if the cigar tips matched Maples' teeth rather than Clark's, better to find that out now than during the defense's case at trial.

When Provonsha brought me the casts of the two men's teeth, I could tell at once that Maples' teeth could not have left those marks in the cigar tips. No matter how I angled and skewed the replicas of the tips, there was simply no way to match the indentations with the teeth; Maples' canines were not the right size, shape, or orientation. Clark's, on the other hand, fit almost like a key in a lock. The lower premolar, in particular, nestled perfectly in that triangular indentation. What's more, the cast of Clark's teeth showed an interesting wear pattern slightly to the right of center: clamping Hav-A-Tampa Jewels between the same teeth year after year had left their mark not just on the wooden cigar tips but on the teeth, too.

I'd already concluded that the cigars had been smoked by the same person; now, I felt certain that person was Roy Lee Clark. But my certainty did not guarantee that my testimony would stand up in court: I was certified as an expert witness in forensic anthropology, including topics such as skeletal identification, skeletal trauma, decomposition, and time since death, but I was not certified as an expert in forensic dentistry. To double-check my conclusions and guarantee there could be no doubt about the credentials of the witness, I suggested Provonsha consult a forensic dentist in San Diego, Norman "Skip" Sperber.

Sperber's credentials were impeccable. He was the chief forensic dentist for the San Diego and Imperial

County medical examiner. Back in the 1970s, he and a partner had created a computer-based system for comparing dental impressions with bite marks. That system—adopted first by the state of California, then by the federal government's National Crime Information Center—systematized dental comparisons in much the same way the FBI's Automated Fingerprint Identification System (AFIS) brought fingerprint comparisons into the computer age. Skip had worked on a number of high-profile murder cases; in perhaps the highest-profile case, he matched bite marks in a murder victim's skin to the teeth of serial killer Ted Bundy.

To my great relief, and that of Provonsha and prosecutor Russ Dedrick, Sperber's analysis—which he conducted without any knowledge of my report—led him to the same conclusion I had reached. It was "highly probable," he said, that the tooth marks in the cigar tips had been made by Roy Lee Clark.

Clark's trial was scheduled for early March 1992. I knew the government was counting on our testimony—mine to lay the foundation of the bite-mark analysis, Sperber's to erect an unshakable, unimpeachable structure on top of it. What I didn't realize was that Sperber and I were the big finish: the final witnesses in the prosecution's case. Even early on, the feds were preparing the jury for what Sperber and I had to say: in Provonsha's testimony about Clark's unsuccessful attempt to steal the Cadillac Eldorado, the FBI agent described how the car's dome light had illuminated Clark's face, revealing a cigar clenched between his teeth . . . on the right-hand side of his mouth. I didn't find out about that piece of testimony until long after the trial was over, but the prosecution did a masterful job of planting a seed in the minds of the jurors.

On the morning of March 5, I arrived in my office at the University of Tennessee to find an urgent message awaiting me: Russ Dedrick's secretary had called; she said Dedrick needed to let me know what time I was scheduled to testify that day. I reported to U.S. District Court right after lunch, and was sworn in as a witness around midafternoon. The prosecutor who would question me on the stand was not Dedrick, however, but Lee Weddle, one of Dedrick's colleagues in the U.S. Attorney's office.

Weddle began, as prosecutors always do, by asking me a long series of questions about my education, experience, and qualifications—not because he didn't know all that already, but so the jury would figure that maybe I really did know what I was talking about. Then he began asking me about the cigar tips.

I explained how we had taken the molds of the cigar tips and compared the replicas with the photos, and then with the casts of Roy Clark's teeth. I also described how Mark Gilbeau and I had measured the angle of the two lower teeth with respect to one another: On a photographic enlargement of the bite marks from the crime scene cigar, we drew a line parallel to the biting edge of the canine, and a line parallel to the biting edge of the first premolar. Those lines intersected, I explained, at an angle of 107 degrees. Then we did the same thing with the bite marks from the cigar butt recovered from Clark's house; that angle also measured 107 degrees.

Next Weddle produced a box containing the casts Betty Crawford had made of Roy Clark's teeth. "Tell me," he said, "whether you received these dental castings for comparison purposes?"

"Yes, sir," I answered. "My name is on the box and also on the cast." I aligned the upper and lower casts so the teeth lined up in what's called the occlusal plane—the orientation Roy Clark's teeth were in as he sat at the defense table—and pointed to the right side of the casts. "The lateral incisor in the upper jaw, which would be this tooth right there, and the canine and the first premolar in the lower jaw are the teeth that made these impressions in this cigar butt," I told the jurors. Several nodded in understanding. "They match up perfectly."

Then Weddle lobbed the key question over the plate, being careful to remind the jury that I wasn't a complete ignoramus: "Based on your education, training, and experience, as you have described [them] to us, and your examination, do you hold an opinion as to whether or not Roy Lee Clark made the teeth marks on those two cigar tips?"

"That is right," I said. "The impressions and the cigar butt found at the scene and in the Clark home were made by Roy Clark."

Clark's attorney, Steven Shope, stood up to cross-examine me. He began by trying to get me to admit that forensic dentistry had long been considered unreliable and unacceptable in court. I told him that I had been testifying about teeth since the 1960s. "I don't know about acceptable," I said, "but they didn't throw me out." That got a laugh from the jurors and a frown from Shope. Then he implied that all thirty-two teeth were needed to identify an individual, or at least more than a mere three. "You could identify an individual from really only one tooth," I responded, "if you had the before-death and after-death record of that tooth."

"Are you saying an individual can be identified with one tooth?" He asked the question with a mixture of disbelief and scorn.

I felt like thanking him for giving me a chance to emphasize my point. "Yes," I said, and explained how one tooth and its relationship to the jawbone was sufficient basis for a positive identification. "Actually," I added, "you can do without *all* of them and *still* make an identification, if your root sockets are good." He would have done well to remember the advice given to every law student in trial class: Never ask a witness a question to which you don't already know the answer.

Shope's final angle of attack was to make the point that wood was not the ideal material for capturing bite marks. "It would be better than glass or steel," I said, "but it's probably not as good as chewing gum or cheese." If only Roy Lee Clark had been considerate enough to leave a half-eaten chunk of cheddar out by the crime scene, and another on the kitchen counter. . . .

After two more questions from Shope, Lee Weddle got his chance to redirect. As before, he ended with a softball question: "Dr. Bass, based on what Mr. Shope has asked you in cross-examination, do you have any reason to change the findings that you reported to this court and jury?"

"No," I said. "I am positive that the teeth on Q65, found at the scene, and the dental impressions found in the Roy Clark house are made by Mr. Clark. I am positive of that."

The next morning, Skip Sperber took the stand. Just as he'd done with me, Weddle led Sperber through a long review of credentials, professional affiliations, and experiences. “I have testified approximately eighty-four times in cases involving bite marks in human skin,” he said, “as well as in other objects such as cheese, chewing gum, things of this nature.” In all, Sperber estimated, he’d been consulted on more than a thousand forensic cases.

Sperber described his examination of the bite marks and related his conclusion that it was “highly probable” that the marks on the two cigar tips came from the same set of teeth. Then Weddle asked him what conclusion he drew when he compared the bite marks with the dental castings from Roy Clark and Richard Maples.

“Mr. Maples’ teeth could not have been responsible for those marks,” said Sperber. “Only Mr. Clark’s could have.”

“No further questions,” said Weddle.

Shortly afterward, the government rested its case.

After two days of deliberations, the jury foreman sent a note to Judge Jarvis saying they were unable to reach a unanimous verdict. Judge Jarvis sent back a note urging them to keep trying. Late the following day, the jury sent out another note. This note informed the judge that they had reached a verdict.

The atmosphere in the courtroom was electric when the jury filed back in. As Clark stood, Judge Jarvis read the verdict: guilty of murder in the first degree. Clark looked stunned and raised his hands, as if in disbelief or horror.

At his sentencing hearing, Clark was offered the opportunity to speak. He seized the occasion to deliver a harangue that rambled on for twenty minutes. In it, he said—among other things—that he had been "crucified"; that when he died and saw Mike Rector "in heaven or hell," he would ask Rector who had killed him; and that Assistant U.S. Attorney Dedrick had "no heart" and was "working for the devil."

After Clark finally wound down, Judge Jarvis pronounced sentence on Clark. In view of the "heinous nature" of the murder, the federal judge sentenced Clark to life in prison without the possibility of parole.

POSTSCRIPT

In 1994, authorities foiled an escape plot by Clark, which involved an attempt to smuggle a gun into the jail cell where he was being held. After that, Clark was moved to Hazelton Penitentiary, a maximum-security federal prison in the mountains of West Virginia. Clark will get out someday, but probably not until he's dead and cold. As dead and cold as Mike Rector was the morning his wife found him facedown in the grass and dandelions, his mail and gun belt beside him, a block away from a well-chewed Hav-A-Tampa cigar butt.

9

LISTENING TO THE BUGS: PINPOINTING TIME SINCE DEATH WITH FORENSIC ENTOMOLOGY

I've spent decades listening to bones—harnessing my academic training, field experience, and mental faculties to hear the stories that bludgeoned, burned, or otherwise damaged skeletal material can whisper. To hear what they have to tell me, I have to pay close attention to the bones and tune out all the background noise. Not surprisingly, that background noise often includes the buzzing of blowflies, the first and most prolific of the critters that come to feed on a human body. In some cases, though, the flies' buzzing isn't background noise to be swatted aside: the buzzing itself is the story, or at least an important part of it, pointing to the manner or timing of someone's murder.

Forensic entomology—finding evidence in insects and insect activity—dates back eight centuries, although that long interval hasn't exactly been a period of rapid or even steady advances. In a remarkable forensic text written in 1247 A.D., a Chinese death investigator named Sung Tz'u told how a murderer was caught by flies. A man in a small village had been murdered, his throat cut. The investigator suspected that the murder weapon was a sickle, a sharp-edged implement used by rice farmers to harvest their crop. Summoning all the men of the village, the investigator ordered them to bring their sickles and place them on the ground. Soon seventy or eighty sickles lay before the assembled crowd. A swarm of flies immediately converged on one of the blades, whereupon the investigator confronted that sickle's owner, charging that his blade bore the blood of the murder victim. Confronted with the entomological evidence—an accurate reflection of the blowfly's keen nose for blood—the man promptly confessed, according to Sung Tz'u.

Seven and a half centuries later and half a world away from that ancient Chinese village, flies would play a pivotal role in case 91-23: the UT Anthropology's twenty-third forensic case of 1991. The case would hinge on flies and fifty yards, fifty yards being the distance that spelled the difference between being burned in Alabama and being burned in Tennessee.

From a murder victim's perspective, of course, fifty yards and a state line are irrelevant: dead is dead and burned is burned, no matter which side of an invisible boundary you're on. From the victim's perspective, the only meaningful boundary is the one between life and death. From the perspective of law enforcement, though, the dividing line between Alabama and Tennessee meant that a Tennessee district attorney picked up the phone and called the Anthropology Department at the University of Tennessee. And *that* meant that my students and I were faced with the challenge of sifting through a set of incinerated human remains in an effort to identify the victim, determine the time since death, and decipher whatever clues had been left behind by the bugs.

One of the Tennessee counties that borders Alabama is Lawrence, a mostly rural county of farmland and forests. The county seat of Lawrence County is Lawrenceburg, a town of about 11,000. One of those 11,000 people, and the one I know best, is District Attorney Mike Bottoms.

On Tuesday, June 4, 1991, Mike called to ask for help recovering and identifying a burned skeleton, which was found in a torched car about fifty yards north of the Tennessee-Alabama border. Early Wednesday morning, two of my graduate students, Theresa Woltanski and Bruce Wayne, loaded up the Anthropology Department's pickup truck and headed southwest for Lawrenceburg, about 275 miles southwest of Knoxville. Normally I would have been eager to go supervise the forensic team, but my wife Ann was battling colon cancer, and I wanted to stay close to home and to the hospital.

At the courthouse in Lawrenceburg, Theresa and Bruce met District Attorney Bottoms and investigators from the Lawrence County Sheriff's Department, the Tennessee Bureau of Investigation, the Tennessee Highway Patrol, and the Tennessee State Fire Marshall's Office. The group's caravan—practically parade-sized for Lawrenceburg—headed south toward Alabama, passing through St. Joseph (population 800) and Iron City (population 350), towns that made Lawrenceburg look immense by comparison.

A couple of miles beyond Iron City, Butler Creek Road curved to the left; just past that, the law enforcement vehicles turned south on a narrow dirt track and bumped south another 150 yards. There, a stone's throw from the Alabama line, was a burned-out Buick Skylark—a four-door 1968 model. The car was gold, or had been; now it was a rusted, charred shell; the windshield had melted from the intense heat of the fire (a telltale sign of arson); the front tires had burned; and the interior had been reduced to rusting metal seat springs and floor pans. The paint had blistered off from the heat of the fire, and both the interior and the exterior were heavily rusted, with rust already starting to flake from the underside of the roof.

The guy in the car had clearly not died at the wheel, or at least had not burned at the wheel: his charred skeletal remains were in the backseat, lying faceup (or what would have been faceup, if he'd still had a face) atop the rusting coils of springs. The head lay on the right side of the seat, and the feet angled down into the footwell behind the driver's seat. The right arm and hand were flopped down into the right rear floorboard.

Theresa and Bruce opened a body bag on the hood of the car to contain the skeletal elements, then began by leaning in through the car's windows. Once they'd plucked out the larger bones, that technique got more difficult, requiring them to crawl halfway through the windows and hang head-down. The car's doors wouldn't open, though, so Scott Brandon, the Tennessee Highway Patrol Investigator, retrieved a pair of bolt cutters from his trunk, and he and the other officers cut the left rear door from the burned car.

The floor pan contained a soupy mix of burned bones, charred debris from the car, and standing water, left over from rainfall during the previous weekend. Theresa and Bruce recovered all the skeletal elements they could find in the muck; the rest, which probably contained smaller fragments of bone, would be sifted by the TBI crime lab and chemically tested for the presence of gasoline or other accelerants.

They brought the skeletal remains back to UT. The first thing we did was to X-ray them. Over the years, I've learned to always X-ray burned bodies, because sometimes they contain evidence that's not visible to the naked eye. In one case, for instance, an X-ray of a burned skull revealed a bright white spot of radio-opaque material in the frontal sinus, above the eyes; that spot turned out to be lead from a bullet, which melted in the fire set by the killer and then trickled into the sinus. If we hadn't X-rayed that skull, we'd never have known the dead man was killed by a gunshot to the head.

The X-rays of 91-23 revealed a bit of radio-opaque material; however, it appeared to be small, random particles, rather than the dense smears or blobs that characterize the lead wipes or melted slugs left by gunshots. Because of their size and random distribution—as well as the condition of the car the remains came from—we judged the particles to be flakes of rust. If 91-23 had been killed by rust particles, this would be a case for the textbooks—maybe even the Guinness Book of World Records.

Although the fire had been hot enough to fracture, shrink, and warp the bones, it hadn't come anywhere near destroying them, as the killer or killers may have hoped it would, and Theresa and Bruce managed to recover most of the skeletal elements. The pelvis was classically male: it exhibited no ventral arc and no concavity in the subpubic region—in other words, it wasn't designed to incubate or deliver a baby. Other distinctive male features included a prominent bump at the base of the skull (the external occipital protuberance) and a large-diameter ball, or head, on the proximal end of the humerus, where it joins the shoulder. The humeral head measured 48 millimeters, or nearly two inches, in diameter—well above the recorded maximum for females, 47 millimeters. Two other distinctive skeletal features that had survived the fire were a well-developed nasal spine (the hallmark of a narrow nose) and a nasal sill, a sort of flange or dam at the base of the nasal opening. Both of these features are characteristic of Caucasoids, or whites.

To estimate age, we used two methods. First we inspected the face of the left pubic symphysis—where the left pubic bone meets the right pubic bone, at the midline of the body—and compared its texture to the reference standards included in a landmark study by anthropologists Tom McKern and T. Dale Stewart in 1957. On the basis of the pubic symphysis, 91-23 appeared to be a young man between twenty-two and twenty-eight years old. Next we looked at the maxillary sutures, the joints in the roof of the mouth. Two of these sutures, which normally fuse during the early thirties, were completely fused or obliterated; a third, which normally doesn't close until the early thirties, showed no signs of obliteration. Combining and averaging out the age ranges indicated by these two methods, we estimated the age at twenty-five to thirty.

Estimating the man's stature was a bit more complicated. Normally the femur (thigh bone) provides the best basis for calculating stature, but neither femur was intact—both were missing the distal portions of their shafts. Fortunately, just a year before, Richard Jantz and one of our Ph.D. students, Tal Simmons, had refined a mathematical formula for calculating stature by measuring the breadth of the proximal (upper) femur. Using this formula, we estimated the stature of 91-23 to be around five feet six inches, plus or minus a couple of inches or so. It wasn't as precise a calculation as we could have done if we'd had an intact femur, but at least it told us we weren't dealing with the skeletal remains of, say, a midget or an NBA basketball center.

The X-rays made it appear unlikely that a gunshot had killed 91-23, but the thermal stresses created by the fire complicated our search for signs of skeletal trauma. Fleshed bone—also called "green bone"—tends to fracture and warp as the moisture inside heats rapidly and vaporizes into steam. Some of the bones were smoked and blackened; others were completely calcined, reduced to their dry, brittle mineral matrix, as if they had been burned in a cremation furnace. The limbs and the cranium, in particular, were badly damaged. The limbs are surrounded by oxygen and are normally the first part of the body to burn; the head has only a thin layer of flesh to protect the bone, and once that burns, the layers of the skull tend to delaminate, or peel apart. Not surprisingly, the skull was highly fragmented and largely calcined; as a result, we couldn't tell whether any cranial trauma had been inflicted before the fire.

Luckily, the bones of the torso—the ribs, the spine, and the pelvis—are better protected, by thicker flesh, and tend to survive fire better than the limbs and the skull do. As we examined the spine, we found two very clear cut marks on the third lumbar vertebra, in the lower torso—a few inches below the protective framework of the rib cage. The larger of the two cuts measured nearly half an inch long, extending across the ventral (front) part of the vertebra. The other cut measured only about a fifth of an inch long on the ventral surface, but it extended deeper into the bone, angling upward and emerging out the upper surface; in other words, it carried enough force to punch through a quarter inch of dense vertebral bone. We also found a large defect on the sternum, or breastbone, which could have been caused by a blow from a sharp instrument, though its cause wasn't as clear as the cut marks in the vertebra, which had obviously been made by a thin, sharp blade at least several inches long—long enough to pierce the belly, transect the abdomen, and slice into the spine.

We had quickly established, despite someone's efforts to destroy the evidence, that the victim was an adult white male, age twenty-five to thirty, of medium height, who had been stabbed at least twice through the abdomen (or once, with two differently angled thrusts within the abdomen from a single penetration wound). He'd possibly been stabbed in the sternum as well, hard enough to break the bone apart, but not hard enough to pierce it entirely.

But who was he, 91-23, and how long ago had he been stabbed to death? To answer the first of those questions, we went straight to the horse's mouth, so to speak: 91-23's teeth and dental work might be able to tell us who he was. Although most of the front teeth had been fragmented by the intense heat of the fire, those farther back in the mouth had been better protected and remained largely intact. We took the skull and mandible to the UT dental clinic, where two dentists—Drs. Wilson and Hurley—made X-rays at the same angles used for living patients. One of the upper premolars, they noticed, had a distinctive "pinch" in the root, and one of the third molars, or wisdom teeth, had an unusual twist in the root. Drs. Wilson and Hurley also noticed that a lower premolar was slightly rotated in the jaw. And then there was the bridge. Amid the burned debris in the car, Theresa and Bruce had found a gold dental bridge—two gold crowns, spanning a gap in the left lower jaw between the second molar and the second premolar.

On June 6, just one day after Theresa and Bruce had excavated the burned car near the Alabama state line, TBI Special Agent Bob Schlafly sent a set of dental records for comparison to the postmortem X-rays we'd taken and the gold bridge found in the car. The dental records documented the teeth and dental work of the man who owned the car: Richard Jamieson, a twenty-eight-year-old who worked for Reed Marine Construction in Florence, Alabama—a riverfront city about four hundred miles downstream from Knoxville. According to Jamieson's employer, Belinda Reed—who filed a missing-person report on May 22—she last saw him on May 17 on the shore of Wilson Lake. He'd been working for the company for about a year, she said, and had never missed work or even been late. He left work at midafternoon on a Friday and didn't show up for work the following Monday, Tuesday, or Wednesday.

Jamieson's dental records came from a dentist in Vernon, Texas, where he lived before moving to Alabama. I don't know how the TBI agent managed to track down the dentist and get the records so fast, but however he did, it was impressive. The records included a set of X-rays, taken in 1984, which showed a pinched root in a premolar, a familiar-looking twist in a wisdom tooth root, and a rotated lower premolar. Tucked in with the X-rays was a note from Dr. Joel Lowry, the dentist, noting that after the X-rays were taken, he had pulled Richard Jamieson's lower left first molar and put in a gold bridge with two gold crowns. So it was official: we could positively identify this stabbed, incinerated murder victim as Richard Jamieson, a twenty-eight-year-old white male. In life, Jamieson was several inches taller than we'd estimated. However, considering that burning can shrink bone by up to 25 percent, we'd gotten pretty darn close, given that all we had to go on was the breadth of the femur.

We knew who our victim was, and we knew how he had died; all that remained for us to figure out was *when* he had died. By the time Richard Jamieson's charred bones were plucked from the hulk of his torched Skylark, we'd conducted a decade of systematic research on human decomposition at the Body Farm. We'd closely observed the changes in the human body during the extended postmortem interval, correlating those changes with both the passage of time and the cumulative effects of temperature. By looking at a bloated or shriveled corpse and then studying temperature records from the National Weather Service, I could tell, with reasonable accuracy and certainty, how many days or weeks had elapsed between the moment someone was killed and the moment their body was found. But that research—and the certainty and the accuracy—was based on nature taking its course with a body. And an accelerant-fueled fire did not count as nature taking its course. The fire that consumed Richard Jamieson's flesh did not conceal his identity, but it did disrupt the normal decomp process.

Paradoxically, as it burned, the fire also froze another piece of evidence we could use to estimate Jamieson's time since death. Within moments of his murder Jamieson had begun to decay; he ceased to smell like a living person, and began, very faintly, to smell like a dead one. Although we humans might not notice the odor of death for hours or days, blowflies can detect it almost instantly. They flock to a body—if they can get to it—and seek out moist orifices (the nose, mouth, eyes, ears, and genitals) and bloody wounds. There the females lay grainy white clumps of eggs which—in late spring, along the Tennessee-Alabama border—can hatch within twenty-four hours or less into thousands upon thousands of wormlike larvae called maggots. Roughly ten to twelve days after hatching, the maggots have grown from tiny specks to nearly an inch in length; at that point, they encase themselves in cocoons called "puparia"—their version of a butterfly's chrysalis—where they undergo a remarkable metamorphosis, transforming from wriggling larvae into winged adult blowflies. The entire cycle, from the time a female lays a clump of eggs until the moment the tiny adult flies chew their way out of the puparia, normally takes around fourteen days—a day or so more in cooler weather, a day or so less in hot weather. By collecting and examining the maggots on a body, it's possible for a forensic entomologist (or an experienced anthropologist) to estimate how far along the larvae are in that fourteen-day development cycle—and therefore how long ago the first blowflies landed on the fresh corpse and laid their eggs. The key to accuracy is to collect the largest, most mature maggots, because they represent the first of the eggs to hatch. I've seen that process, the blowfly's predictable and consistent life cycle, unfold countless times by now. But until 1981, I'd never had a chance to observe it systematically, in a steady stream of human corpses. No one had.

I don't mean to imply that no one had conducted studies or made forensic use of the life cycle of flies and maggots. During the 1800s, physicians and scientists in France and Germany made noteworthy strides in documenting and forensically interpreting insect activity in corpses. In America, a major entomological study was conducted in the 1890s by Murray Galt Motter, who examined the numerous bugs that came to light when a cemetery in Washington, D.C., was moved, requiring the exhumation and relocation of some 150 bodies.

But Motter's research focused on bodies that had been sealed in coffins and buried in a cemetery for years, not bodies recently tossed in the woods or burned in cars; as a result, it wasn't terribly useful in forensic cases, at least not for determining time since death. The creation of the Body Farm gave us an unprecedented opportunity to fill that research gap. In fact, one of the earliest and most influential research projects at the Body Farm—one that began as soon as we'd completed the research enclosure and obtained our first research body in 1981 (body 1-81)—was a study by graduate student Bill Rodriguez. Over the course of many months and several bodies, Bill documented which insects fed upon corpses, and when each species arrived at the feast. That study, which created a sensation at the conference where Bill presented his results (complete with graphic slides of the insects' feeding frenzy), became a landmark reference in the development of modern forensic entomology. Ever since, crime scene investigators—including my graduate students—have known to collect maggot specimens off bodies, and to divide them into two batches. One batch should be placed immediately in a "kill jar"—a container with enough ethyl acetate or alcohol fumes to kill the bugs—to halt their development. The other batch should be brought back to the lab and allowed to complete their development, to see how much longer it takes to complete the fourteen-day life cycle.

Theresa and Bruce knew all this, and arrived at the scene equipped to bring back maggots both dead and alive. Theresa and Bruce dutifully collected the fattest of the maggots that had hatched from the blowfly eggs. They also collected a number of small cylindrical objects they'd been trained to look for: the puparia in which maggots encase themselves to undergo their metamorphosis into flies. The mere presence of the puparia was significant, because they told us that somewhere around fourteen days had already elapsed between the murder and the time the body was found—possibly as few as ten or twelve, given the summery weather that had already engulfed southern Tennessee and northern Alabama by late May and early June.

As we studied the puparia, though, we noticed something unusual, something I'd never seen in any other case. The puparia from the car were black. Actually, that's not precisely true; they weren't black, but blackened—charred—just like the car itself, and just like Richard Jamieson's bones. The implications were startling: Jamieson had been dead and rotting in the backseat of his car for the complete duration of the blowfly life cycle *before* his car was torched. His killer or killers, in other words, had hidden the car and the body in the woods, gone away for nearly two weeks, then come back, doused the vehicle and the body with gasoline, and set it ablaze.

Neither the TBI nor the Alabama authorities investigating the murder ever managed to arrest a suspect. Richard Jamieson may have been a model employee at Reed Marine Construction, but he was not a model citizen; in his off-hours, the TBI found, he was apparently a small-time drug dealer and a bigger-time drug user, with a penchant for owing money to unsavory characters. Before moving to Alabama, he'd reportedly gotten deeply enough in debt, to sufficiently unsavory folks, to receive death threats. It wasn't hard to imagine the same thing happening in Alabama, with the ultimate payment coming due before Jamieson had a chance to pay up or relocate again. That's just a theory, mind you, but it fits the other evidence.

Regardless of the motive for the murder, whoever killed Jamieson made a concerted, incendiary effort to destroy evidence of Jamieson's identity and manner of death. They came pretty close to succeeding, too. They only missed it by a scant fifty yards and a thick swarm of flies.

10

THE PROFESSOR VERSUS THE COMPUTER: HARNESSING SOFTWARE TO DETERMINE RACE AND ID A VICTIM

For virtually my entire half-century career, and for many decades before I began my studies in osteology (bones) and anthropology, the skull was considered the only reliable basis for determining the race of an unknown victim's skeleton. In fact, as recently as 2006, I retained this passage in a new, updated edition of my guide to bones, *Human Osteology: A Laboratory and Field Manual*: "The skull is the only area of the skeleton from which an accurate estimation of racial origin may be obtained."

In hindsight, I think I should have reworded that passage, made it less sweeping, because techniques for determining race are changing. One reason for the change stems from research by one of my former graduate students, Dr. Emily Craig, who now serves as the forensic anthropologist in the state medical examiner's office in Kentucky. For her Ph.D. dissertation, Emily X-rayed hundreds of femora (femur bones) and found that "Blumensatt's line"—the angle at which the shaft of the femur meets the condyles (the femur's broad, load-bearing surfaces at the knee)—is consistently different in blacks than in whites. Another of our graduate students, Jennifer Synstelien, did a master's degree thesis several years ago that demonstrated that innominates—hip bones—also exhibit consistent racial differences, being generally smaller and shorter in blacks than in whites.

Another enormous factor in our growing ability to determine race from a variety of bones is the Forensic Anthropology Data Bank, created at UT in 1986, whose measurements of thousands of individual human skeletons provide an ever more detailed picture both of ancient racial differences and of rapid modern changes. The Forensic Data Bank allows us to detect subtleties we hadn't known existed. It also provides the foundation for a powerful new computer tool called ForDisc, which can analyze those subtleties in milliseconds to confirm what a seasoned anthropologist already knows—or to challenge that anthropologist's experience and intuition with cold, hard numbers.

One afternoon in the fall of 1991, I got a call from the Roane County Sheriff's Office saying that a bundle of plastic containing human skeletal remains had just been found in Polecat Creek, which fed into the Tennessee River near the boundary of Roane and Loudon counties. I wasn't there to take the call; my wife Ann was losing her battle with cancer, and I was spending as much time as I could with her. But by this time—twenty years after I'd arrived in Knoxville and a decade after we'd created the Anthropology Research Facility—we had a well-established routine for handling forensic cases, whether I was in town or on the other side of the globe.

Three of my graduate students were on call this particular November day: Bill Grant, Theresa Woltanski (who later married Bill), and Juliet Cleaves. They climbed into the cab of the Anthropology Department's pickup truck and headed south on 1–75 to exit 72, where they took River Road to the bridge over Polecat Creek. This was our forty-second forensic case of 1991; that meant the victim would be designated, until we could discover his or her identity, simply as 91-42.

They got there at about two-fifteen. By that time, divers had already cut an opening in the plastic and recovered a pair of shoes and socks, which contained the bones of both feet, and a sacrum—the wedge-shaped lower end of the spine, consisting of five fused vertebrae.

The law enforcement officers had been steered to the spot by an anonymous caller. The caller described seeing a pregnant woman being shot and thrown into the creek from the bridge; the woman had been missing for ten months.

As soon as Bill saw the sacrum, he told the officers, "That looks like a male sacrum." The officers corrected him, pointing out that they were searching for a female, and that the caller had specifically described the missing pregnant woman. Bill repeated his belief that the sacrum appeared to be male—it was quite robust—but the officers were confident they had found the missing woman. The shoes were in such bad shape they offered no help in resolving the argument; the bones in the socks could have been a man's or a woman's.

In addition to these loose items, the divers had brought up two pieces of adipocere—fatty tissue that had been transformed, by chemical reactions in the water, into a waxy or soapy substance. All these came from a large bundle wrapped in heavy black plastic, which they'd slit open under the water. The bundle was fastened with wire cable to five cinder blocks. When someone drowns, or when a body is thrown into a lake or river, the corpse often sinks initially. Eventually, though, as the body's own acids and enzymes begin to digest it from the inside out, the process of decomposition creates gases as a by-product. During the "bloat" stage of decomposition, these gases build up in the abdomen, causing it to swell . . . and transforming a submerged body into what is commonly called a "floater." In earlier times, people believed that firing a cannon near the site of a drowning could cause the victim's body to rise to the surface. And it could—if the cannon shot just happened to coincide with the precise moment when the buildup of gases created positive buoyancy in the corpse. But weighted down by nearly two hundred pounds of cinder blocks, the body tossed into Polecat Creek would have required a buoyancy nearly equal to the Goodyear blimp's to float to the surface. Clearly someone who knew something about decomposition wanted to make sure this body never became a floater.

As the divers described the trussed-up plastic bundle resting on the bottom of Polecat Creek, the students stressed the importance of not losing any of the skeletal material as the bundle of plastic was hoisted. To minimize the risk of tearing the plastic and scattering bones in the muck of the creek bed, two of the divers lowered a metal, basket-style litter into the creek. Next they cut the cables that held the bundle to the cinder blocks, eased the bundle into the litter, and brought it carefully to the bank. The bundle was held together by more of the cable wire that had tethered it to the blocks, so after the team took photos, they cut the wires and slit open the plastic. Inside was a nearly complete skeleton; upon inspecting the skull and pelvis, all three of the graduate students, who had extensive experience in identifying bones, told the officers that the skeleton appeared to be male. Again—despite the graduate students' expertise in osteology training (and the lack of a fetal skeleton)—the officers insisted that the victim was a female.

To complete a forensic examination, the team needed to bring the remains back to the UT Medical Center and the morgue so the corpse could be X-rayed (in case there was a bullet or traces of lead somewhere) and the soft tissue removed from the bones. The bundle of plastic (which appeared to be black polyethylene, of the type used in landscaping) was carefully zipped inside a body bag, stowed in the back of the truck inside the camper shell, and brought back to the Anthropology Department. We would examine them the following day.

First thing the next morning, we set to work. The remains were almost completely skeletonized, which made them easier to clean and study. The skull still contained part of the brain, and adipocere clung to some of the bones, but it was far less messy than the dozens of floaters I'd examined over the years, after they'd surfaced in one of Tennessee's numerous lakes and rivers.

Two of the Roane County officers arrived, still convinced that these bones belonged to a white female. Well, they were half right, I said; the skull was white, but there was no doubt that it was male, not female. The brow ridge, above the eyes, was pronounced and heavy. The external occipital protuberance—the bump at the base of the skull where several neck muscles attach—was prominent as well, another distinguishing male trait. The pelvis was also unambiguously male: for starters, the pubic bones were far too narrow to allow a baby's head to pass through; in addition, the rest of the pelvic geometry was consistent with the narrow pelvis of a male. So was the size of the long bones of the upper arm and the thigh: the head of the humerus (where the arm joins the shoulder) measured 51 millimeters (two inches) in diameter, while females virtually never have a humeral head larger than 45 millimeters; the head of the femur—the "ball" that rotates in the hip joint's cup-shaped socket—measured 48 millimeters in diameter, which put it into the male range as well. The femur's length could also help us estimate the victim's stature. After measuring hundreds of femur bones from anatomical specimens of known height, anthropologist Mildred Trotter had developed a formula for extrapolating an unknown person's stature from a single measurement of the femur. The length of 91-42's femur was 52.3 centimeters (20.6 inches). Plugging that number into Trotter's formula (length × 2.11 × 70.35) told us that in life, 91-42 had stood 180.7 centimeters tall, or 5 feet 11 inches.

The shape and size of the bones all added up to one consistent, overwhelming conclusion: I was 100 percent certain 91-42 was male, I told the officers. To underscore the point, I brought out other male skulls and pelvises from the skeletal collection, as well as female skulls and pelvises, and showed them the obvious differences. Reluctantly, they finally accepted the fact that they hadn't found the pregnant female they were looking for.

To estimate the man's age, we first examined the pubic symphysis, the joint at the midline of the body where the left pubic bone connected to the right pubic bone. Using a method devised by Tom McKern and T. Dale Stewart in 1957, based on a study of 349 pubic symphyses from males of known age, we looked at three parts of the pubic symphysis from 91-42: the dorsal demiface, the ventral rampart, and the symphyseal ring. The dorsal demiface is the roughly semicircular face of the symphysis that's closer to the spine. In 91-42, the dorsal demiface showed pronounced degeneration—the sort of wear that occurs during several decades of aging. The ventral rampart is the symphyseal surface that's closer to the front of the abdomen; in 91-42, the ventral rampart, too, was extremely worn, as was the symphyseal ring, the raised ridge around the circumference of the symphysis. According to McKern and Stewart's method, this amount of wear corresponds to an age of at least thirty-six years. However, I didn't have complete faith in the accuracy of that age estimate, because both of the pubic bones were somewhat eroded from their long immersion in water. For that reason, we also looked at the auricular surface of the ilium—the large, ear-shaped portion of the innominate, or hip bone. The ilium was less eroded than the pubic symphysis, and showed classic signs of middle age: the surface looked extremely dense and smooth, having lost most of the micropores, striations, and granules that characterize this surface in a younger person. These textural features helped narrow our victim's age to somewhere between forty and forty-nine.

Determining what had killed him came next. Over the years, I'd learned always to X-ray burned bodies and bodies recovered from water: killers often tried to cover their tracks by burning their victim in a house fire or a car fire, or by dumping them in a river, so in a fairly high percentage of such bodies, we found bullets or lead smears. In this case, of course—barring the world's most methodical and dogged suicide, someone who took the trouble to wrap himself in plastic and tie himself to five cinder blocks before somehow surmounting the bridge railing—we already knew a murder had been committed. What we needed to confirm was how it had been committed. To find out, we took radiographs of 91-42's head, chest, and pelvic regions, using a portable X-ray machine in the morgue at UT Medical Center (not surprisingly, the hospital's Radiology Department frowned on having slimy, smelly corpses brought into the same area where patients were X-rayed). When the films were developed, the manner of death was obvious: in and around the skull were several smears of bright white—material that was too dense for X-rays to pass through. The smears were lead from one or more bullets; 41-92 had been shot in the head, maybe more than once.

When we cleaned the skull, we found the copper jacket from a bullet embedded in the brain. We also found what appeared to be a copper jacket within the leather coat that the skeleton was still "wearing." The skull was heavily damaged. The left side of the maxilla, or upper jaw, had a semicircular fracture and was highly fragmented, and most of the palate was missing.

So was the floor of the left eye orbit (socket); the top of the orbit had a linear fracture running through it. The two upper left incisors were broken away, with two small cracks extending along the top of the maxillary plate of bone. This pattern of damage indicated that the bullet had entered the skull through the left cheek, probably right beside the front teeth.

Once it had blasted through the face, the bullet continued through the skull, angling slightly downward and to the right. It struck the occipital bone, which forms the base of the skull, in the area of the foramen magnum, the large opening through which the spinal cord exits, and also hit the first cervical vertebra, fracturing it. The bullet's impact caused multiple fractures in the rear of the skull, as well as knocking loose a chunk of bone around the foramen magnum and another chunk farther back, at the right rear base of the skull. At some point, as it ricocheted inside the skull, the bullet split; we found one fragment stuck in the soft tissue of the scalp, on the exterior surface of the occipital, while other fragments remained inside the cranium.

While the cause of death was easy to determine, the murder victim's race was not. Some of his skull's features were characteristic of black people; other features, of whites.

In recent years, the concept of race has come under intense scrutiny and even severe criticism, even—maybe especially—from other anthropologists. Most cultural anthropologists today regard race as a cultural or social construct, not a biological distinction. I'm not a cultural anthropologist, though, I'm a physical anthropologist; over the past half century, I've studied thousands of skulls, and I can say with great certainty that there are clear and consistent differences between the skeletons (especially the skulls) of people in the races—or the groups, or whatever label you prefer to apply—that anthropologists a century ago labeled Caucasoid, Negroid, and Mongoloid.

Even within those broad racial groups, there are differences. For example, when I was a graduate student in Philadelphia, studying under Dr. Wilton Krogman, the leading "bone detective" of his time, a man was struck and killed by a subway train. Actually, within the city, they're subways; farther out, the trains run aboveground. This death occurred out in the suburbs one snowy winter's day. The motorman was sitting up in the front of the lead car, and he saw a man coming down the embankment alongside the tracks. The man was wearing an overcoat and had the collar pulled up over his ears to protect them from the bitter cold. The motorman began blowing the horn, but the guy apparently never saw or heard the train coming; or if he did, he was suicidal, because he stepped right into its path. It was the neatest, cleanest coronal dissection I ever saw: it took the first three inches off the front of the man's head—his face and forehead—as well as his right arm, which had swung in front of his body as he took his last stride. The front part of his head was on the front of the subway car; we scraped it off and took it back to the lab, along with the rest of his body, which lay alongside the tracks.

The man's skin was extremely dark, and that was unusual for Philadelphia at the time—most black people in Philly had lighter skin, more of a caramel or mocha color. To understand why, you need to know a bit about history—specifically, the history of slavery in America.

Back in the Deep South hundreds of years ago, plantation owners were picky about the Africans they purchased as slaves. Slaves were expected to work hardest during the hottest part of the summer, when tobacco crops and cotton fields needed planting and weeding. Consequently, the plantation owners preferred to buy Africans who had come from the hottest parts of Africa. It didn't take too many shiploads of slaves for the planters to notice that the most heat-tolerant Africans consistently had gums that were a dark bluish-black in color. These people came from the western coast of sub-Saharan Africa, and they were nicknamed "bluegums." And in the free-market economy of supply and demand that treated human beings as commodities to be bought and sold, bluegums brought higher prices on the auction blocks of the Deep South, because they could withstand heat better.

A century after the Civil War brought an end to slavery, as Dr. Krogman and I scraped a face off a subway car and studied it for clues to the man's identity, the aftereffects of slave-trade economics were still visible. "This guy had blue gums," Dr. Krogman said, and launched into an explanation of the significance. "I bet you this guy is from south Alabama or Mississippi; he probably came North looking for work, and he probably doesn't write home." I don't know if Krogman's entire hypothesis was correct, because the man was never identified, but he was right about one thing, I noticed in the weeks and months that followed: our subway casualty did indeed have far darker skin than most blacks in Philadelphia.

For millennia, racial distinctions corresponded with geographic divisions; that is, most Negroid people were in Africa, most Caucasoids in Europe, and most Mongoloids in Asia. Today, though, jet aircraft and bullet trains have rendered geographic barriers virtually meaningless. At the same time, cultural barriers against racial mixing have been toppling as well. As a result, the distinction between racial groups is lessening. If this trend continues—and I see no reason why it won't, nor why it shouldn't—eventually the human races will tend toward one human race, with a common, shared set of skeletal characteristics. In the meantime, though, it's helpful to be able to tell police whether an unknown murder victim is white, black, Asian, or Native American.

It's helpful, but it's not always easy. Our plastic-shrouded gunshot victim, 91-42, was racially ambiguous. He possessed some Negroid features, including a long, low skull, wide-set eyes, and a short nasal opening. But he also exhibited Caucasoid features, including a "sill" at the base of the nasal opening, a sharp nasal spine (the bone that forms the bridge of the nose), and a hooked inion (the point on the end of the external occipital protuberance, which is the bump men have at the base of their skull).

One logical explanation for this ambiguity could have been racial mixing. If a black male and a white female (or vice versa) breed and produce ten children, each of the ten will inherit a slightly different combination of racial features or traits. Some will inherit broad noses, others will have narrow noses; some will have thick lips, others will have thin lips. Skin coloring will vary, too. It's not like pouring red food coloring into water, which creates a uniform pink liquid; the results will be different every time. If just the noses changed, for instance, determining race (or racial mixing) wouldn't be that difficult, but since everything changes, it's fluid.

In the case of 91-42, then, it was certainly possible that one of his parents was white and the other black, but I didn't think that was likely. After all, he had been born somewhere between 1942 and 1951, and back then interracial couples were rare, at least in the South. It was a tough call, but on balance, my hunch was that our victim was white.

Captain Moldenhauer searched his memory for any missing-person reports describing a white male age forty to forty-nine. He quickly drew a blank.

Had I been wrong? If Dr. Krogman were still around, I'd have consulted him about the skull, but Krogman had died four years before. So Instead of consulting my mentor, I consulted a protégé, Richard Jantz. Richard, in turn, consulted *his* protégé—a new computer program he'd just developed.

I had met Richard three decades earlier, when he took my introductory anthropology course at the University of Kansas. Dick was a Kansas boy, born and bred; his father worked at a grain elevator in Halstead, a little prairie town about thirty miles northwest of Wichita. Dick was the oldest of three boys; he and his younger brother both attended college in Lawrence, and I taught both at one time or another. Dick got interested in anthropology and took a number of undergraduate courses; eventually he decided to get a Ph.D., and I ended up as his thesis advisor. Like me, Dick found bones more interesting than cultural rituals and customs; unlike me, he also loved statistics and data. During graduate school, Dick spent a summer with me in the plains of South Dakota excavating Arikara Indian graves that were about to be inundated by the rising waters of Lake Oahe. (For the story of how I became "Indian grave robber number one," see *Death's Acre.*) In the course of that summer, Dick catalogued and measured thousands of Indian bones. He wrote his dissertation on cranial changes in the Arikara Indians between 1600 and 1830, and it was a superb piece of work. When I came to Tennessee in 1971 to head UT's Anthropology Department, Dick—who was an assistant professor at the University of Nebraska at the time—was one of the two bright young faculty members I'd negotiated to bring with me.

Plant enough seeds, and sooner or later something's going to take root. Eventually all those bone measurements took root in Dick's numerically inclined brain, and as we entered the computer age, Dick got interested in using computer software to analyze skeletal measurements and provide new ways of determining the race, sex, and stature of unknown skeletons. Dick wasn't the first to think about approaching these questions from a rigorous statistical standpoint; an English scientist named R. A. Fisher had invented the statistical tools years before, but the complex and seemingly endless calculations made the system virtually unusable. During the 1960s, anthropologists Eugene Giles and Orville Elliott devised and published ways to make those more usable. By the 1980s, the revolution in personal computing—steadily shrinking computers packing ever-faster processors—made it possible to perform, in minutes, calculations that would have taken days or weeks for Giles and Elliott—or other scientists using their technique—to perform.

Dick enlisted one of his own graduate students, Steve Ousley, to help write a computer program that would be able to calculate the probable sex, race, and stature of an unknown skeleton, by comparing that unknown's skeletal measurements with those in a database that contained thousands upon thousands of other measurements from skeletons whose sex, race, and stature were known. Initially the measurements in the database came from the skeletal collection we were building from bodies that came through the Body Farm. Then Dick and I—along with graduate assistant Bill Rodriguez—visited other modern skeletal collections to expand the database. The real turning point came when Dick began receiving skeletal measurements from anthropologists all over the world, so that the software would not be limited by the narrow cross section of humans we continued to collect in Knoxville.

After adding an easy-to-use interface, so nonstatisticians could readily run the software, Dick named the program ForDisc, short for "forensic discriminant function analysis," a name that doesn't roll quite as trippingly off the tongue as ForDisc. It took a couple of years to create and refine ForDisc; Dick spent the fall of 1990 and the spring of 1991 on sabbatical in Germany, where he wrote the lion's share of the software. So ForDisc 1.0 was basically finished but as yet untested in November of 1991, when divers retrieved a bundle of plastic-wrapped human remains from the muck at the bottom of Polecat Creek.

When I called Dick into my office, he was intrigued by the case, and thought it would be an interesting test for ForDisc. He took the skull and long bones back to his office; pulling out calipers and osteometric measuring boards, he quickly measured key landmarks on the skull and the lengths, diameters, and other dimensions of the tibia, femur, and humerus. When he plugged those numbers into ForDisc, the software compared the numbers with the database and rendered its analysis. After comparison to thousands of other measurements from hundreds of other skeletons, the computer begged to differ with me: 91-42 was more likely black than white, according to ForDisc. Like me, the software found ambiguities in the skull—initially ForDisc gave a 55 percent probability that 91-42 was black. Looking just at the measurements of the postcranial skeleton, though—the bones from the neck down—ForDisc was far more confident in its prediction, assigning a probability of 95 percent that our victim was black. Performing two different calculations based on measurements from four bones (the clavicle, the radius, the humerus, and the tibia), ForDisc estimated his stature to be six feet—an inch taller than I'd estimated.

Armed with this new data, Moldenhauer left and refocused his search. Instead of trying to pin down the identity of a white female, or even a white male, he was now looking for a black man, one who'd been missing for at least two years, maybe considerably longer.

It didn't take long to find someone who fit the description. In March of 1980, a Lenoir City man named William Austin Goley had disappeared. Goley, a black male, was forty-one at the time of his disappearance.

But was this Goley? The body was completely skeletonized, so there were no fingerprints. DNA testing was still in its infancy, and the odds of recovering usable DNA from bones that had been immersed in bacteria-laden waters for more than eleven years seemed slim. We couldn't find dental records for Goley, to compare against the teeth in the skull (the ones that hadn't been broken off by the gunshot). As it turned out, though, none of that mattered, because we had a right knuckle and a left wrist.

In addition to the massive cranial trauma caused by the gunshot that killed him, 91-42 had two distinguishing skeletal features, both of which predated his death. One was a healed fracture in his right hand, at the base of his little finger. If you close your hand into a fist and you hit a person really hard, you're likely to fracture one or more of the bones that fan out through your hand. Those bones are the metacarpals, and this type of metacarpal fracture is called a "boxer's fracture," since the most common way of getting one is to punch someone really hard. In the head of his fifth (pinkie) metacarpal, 91-42 had a healed boxer's fracture. According to William Austin Goley's medical records, he was diagnosed with just such a fracture in the summer of 1976, after "he did strike another individual or an object."

But far more noticeable than the healed boxer's fracture was the wrist. Goley had been a patient at Knoxville Orthopaedic Clinic between 1971 and 1978; moreover, he'd been a very frequent patient. His medical record for that period ran to nearly fifty pages, containing a litany of injuries and complaints, including back pain, knee pain, soreness in his right wrist, pain in his left hip, and the boxer's fracture in his right hand. One of his many doctor's visits occurred in June of 1974, when he came into the clinic complaining of pain after "his wife hit him with a figurine in his left wrist." After two years, multiple visits, and repeated treatment for inflammation and dislocation of the wrist, Goley had a damaged scaphoid bone removed, and an artificial one put in its place—a prosthesis made of Silastic (silicone plastic). Our victim, 91-42, had a Silastic scaphoid implant. Considering the astronomical odds against two forty-something-year-old black males in Loudon County having a boxer's fracture of the right fifth metacarpal and also a Silastic scaphoid implant in the left wrist, I felt confident making a positive identification: UT forensic case 91-42 was definitely William Austin Goley, age forty-one, height 5 feet 11 inches.

I also felt confident that ForDisc, which had made such an impressive debut in its first real-world case, would prove to be a valuable new tool in forensic anthropology. I was right about that, even if I wasn't initially right about the race of 91-42. In several subsequent U.S. murder cases that I know of, ForDisc has played a pivotal role in focusing or refocusing efforts to identify unknown victims. It has also proved extremely helpful to anthropologists excavating mass graves in Kosovo, Argentina, and other countries where authoritarian governments or ethnic-cleansing movements have massacred people in droves.

But racial mixing complicates things for ForDisc, just as it does for anthropologists. "Things are getting more complicated fast," concedes Dick, "a lot faster than we're figuring them out." American blacks are a hybrid group, as are Hispanics—in fact, Hispanics are even more complicated than blacks to identify by skeletal features because they tend to represent centuries of mingling of Native American traits with European traits. Confronted with such mixed skeletal signals, ForDisc produces ambiguous results and lower probabilities. The software doesn't produce a flashing sign that shouts "hybrid," but if you listen closely to the ambiguities, says Dick, "ForDisc whispers that in your ear."

As the world, and the murders and the victims in it, become more complex, those whispers can become harder to hear and decipher. But the Forensic Data Bank, which ForDisc consults with each new case, is huge by now, and getting bigger all the time. Dick has done a superb job of getting cranial metrics from all over the world—measurements from the mass graves in Kosovo, from elsewhere in Europe, from Asia, and even from ancient archaeological sites in the Nile Valley. I recently heard about an osteology Ph.D. student on a dig in Egypt who was puzzling over an unusual-looking female skull the crew had unearthed. She ran the skull's measurements through ForDisc and found out why it looked different: it wasn't an Egyptian's skull at all, but a young Central European woman, who probably came to Egypt on a ship from Greece.

More and more anthropologists around the world are recognizing the importance of expanding ForDisc's database. Still, ForDisc inevitably lags one step behind the world's living, breathing, blending peoples—you do have to be dead, after all, to have your skeletal measurements taken and entered into the database. Nevertheless, ForDisc grows more sophisticated and powerful with each new version, and we're now at version 3.0. That version—a $395 program that can be downloaded with a credit card, or shipped on a tiny flash drive—is in use around the world, at organizations as varied as the Smithsonian Institution, the U.S. Army Central Identification Laboratory, the Georgia Bureau of Investigation, the FBI, the Royal Canadian Mounted Police, and anthropology departments and medical examiners' offices throughout North America.

As a forensic tool, ForDisc is getting better and better. And the better the tools we have, the better we'll be able to hear the dead whisper in our ears, and the more likely we'll be to hear their stories, answer their pleas, and bring their killers to justice. Like William Austin Goley's killer—a romantic rival—who pled guilty to the crime in 1994, three years after ForDisc whispered "black male."

11

SCIENCE AT THE CUTTING EDGE: SCANNING A KNIFE MARK WITH ELECTRON MICROSCOPY

One steady trend in science—virtually all science, not just forensic science—is the development of new tools and techniques for looking ever closer and closer at ever-smaller things, whether those "things" are biochemical reactions occurring in the brain as it processes various images or sensations, viruses that threaten to kill millions of people, or microscopic fractures or cut marks in the bones of murder victims.

I remember the murders that got solved and successfully prosecuted. I'm haunted by those than remain unsolved. Some haunt me more than others. None haunts me more than Letha Rutherford's. In her case, a powerful modern technology—electron microscopy—gave us remarkable insight, literally, into her manner of death. But insight, sadly, can be a far cry from justice.

In August of 1992, I got a phone call from Brenda Wilson, the mother of an eighteen-year-old girl named Letha Rutherford. On December 16, 1991, Letha spent the day working on a farm stripping tobacco (pulling the leaves off the stalks). At the end of the workday, she caught a ride home with some coworkers, who dropped her off at her house. She wasn't seen again until four months later, on April 14, 1992, when her body, decomposed and partially skeletonized, was found beneath some roofing tin and other debris in a trash heap, a stone's throw behind the small frame house she shared with a friend. Letha had grown up in the house; it was quite possible that she had also died in it. The body was found by Jimmy Hager and his son James, both cousins of Letha's, while James was riding an all-terrain vehicle in the woods behind the house. Ironically, police had searched the area the previous December, on horseback and with dogs, but found nothing. The discovery of the body, behind an isolated house on a dead-end road, prompted Letha's brother, Maurice Rutherford, to speculate that she was killed by someone who knew her. "Who else would go down there?" he asked a Lexington newspaper reporter at the time.

One possible answer to his question came three months after the body was found: in July, police pulled up and burned nearly 1000 marijuana plants growing on 100 acres of irrigated and fertilized land off Dry Branch Road. The marijuana crop, which police estimate would have been worth almost a million dollars, was the largest pot seizure in or near Lexington in five years. According to Letha's mother, Letha suspected a relative of growing the plants, and had threatened to call the police.

After Letha's remains were recovered, the state's deputy medical examiner, Dr. John Hunsaker, turned them over to the state's forensic anthropologist, Dr. David Wolf, in hopes that Wolf might be able to find some telltale evidence amid the bones, clues to how she had been killed. Unfortunately, Dr. Wolf was gravely ill at the time; he died without ever being able to conduct a detailed examination of the skeletal material.

The Kentucky medical examiner's office ruled her death a homicide of undetermined nature and retained some material from the remains, including the pelvis, a few ribs, and the neck, including the hyoid (a delicate arched bone in the throat, which is often fractured in cases of strangulation). On April 29, the rest of Letha's skeletal remains were buried. Her mother scraped together the money for an inexpensive burial. Letha was interred in a cloth-covered coffin, with no headstone to mark her grave.

Three months after the burial, Letha's mother saw a television show (an episode of ABC's *48 Hours*) that featured my work in forensic anthropology. That's what prompted her call to me in August, which she followed up with a letter, pleading for my help in determining how Letha had been killed. Her daughter's case had fallen through the cracks of the justice system, Mrs. Wilson believed, because they were poor. Maybe she was right, or maybe not; in any case, she was understandably distraught about Letha's unsolved murder, and I agreed to help if I could.

Mrs. Wilson had struggled to raise the money to bury Letha; now, she faced a struggle to raise the money to un-bury her. An exhumation would cost close to $700, and for Brenda Wilson, $700 was a lot of money.

But I wasn't the only one moved by this grieving mother's plight. Rob Kaiser, a reporter for the *Lexington Herald-Leader*, knew about the case, and wrote a poignant article that ran on the paper's front page. Beneath a large photo of Brenda Wilson standing beside her daughter's grave, the headline read, COST KEEPS MOTHER FROM ANSWERS. Within days after the story ran, an anonymous donor made a cash contribution to the cemetery to pay for the exhumation. That was in mid-November. A month later—a year and two days after she disappeared—Letha Rutherford's grave was dug up, and her waterlogged coffin was taken to the Kentucky medical examiner's office in Frankfort.

Accompanied by one of my graduate assistants, Murray Marks (now on the anthropology faculty at UT), I drove to Frankfort the morning of Monday, December 21. Our first stop was the Lexington–Fayette County Urban Police Department, where we met with Detective Larry Etherington, the lead investigator, and Sergeant Dan Gibbons. They let us examine the crime scene photos, which showed Letha Rutherford lying facedown wearing only a yellow T-shirt pulled up around her neck. Her legs were apart, with her right leg aligned more or less with her spine and her left leg angled up roughly 90 degrees to the side. Most of the bones were exposed, although soft tissue remained on her lower back and hips. The police had also taken a video documenting the recovery of the body; although I always prefer the opportunity to examine a crime scene in person, in this case—when that clearly wasn't possible—the police department's crime scene images were excellent. That afternoon, we met with Dr. John Hunsaker, the state's associate chief medical examiner. Dr. Hunsaker, Murray, and I looked at the skeletal material and discussed the soft tissue that the M.E.'s office had retained, and the three of us made a plan for cleaning and examining the exhumed remains the following day.

The next morning, Murray and I met Dr. Hunsaker at the M.E.'s forensic laboratory, where we were joined by David Jones, the director of the state M.E.'s office. Also on hand were an officer from the Lexington–Fayette County Urban Police Department, who came to videotape and photograph the examination; Dr. George Nichols, the state's chief medical examiner and forensic pathologist; several members of Dr. Hunsaker's and Nichols' staffs; and Dr. Mary Powell, an anthropologist and the curator of human osteology at the University of Kentucky Museum.

At 9 A.M., as we all gathered around, Letha Rutherford's coffin was opened. Inside was a sealed body bag, which we removed and unzipped. Then everyone but the police videographer pitched in to help sort and clean the remains, starting with the skull and working our way down.

It was not a pleasant task. The remains had been badly decayed when they were found, and seven months underground, sealed in a body bag, had only furthered the putrefaction. A moderate amount of soft tissue remained on the bones; it was slimy, smelly work to scrub the bones under warm water and lay them out in their proper anatomical position, but with all the help, it took us only a couple of hours. Luckily, considering my line of work, I have a poor sense of smell, so the processing was more bearable for me than for the others. While most of us scrubbed bones, Dr. Hunsaker carefully dissected the patch of skin and tissue surrounding the hyoid bone, which the M.E.'s office had retained for the past year so it could be examined for fractures.

When we were done, all the bones were there. Some showed evidence of gnawing by rodents and other carnivores, but amazingly, none had been dragged or scattered during the months the body had lain in the trash heap.

The M.E.'s office had X-rayed the remains right after they were found, but nothing that pointed to a cause of death showed up in either the X-rays or the cursory examination Dr. Wolf had made shortly before his death. I had reviewed the X-rays, too, and could see nothing out of the ordinary on the films. But as soon as I saw one of the bones, I felt my adrenaline spike, and I had a hunch what had killed Letha Rutherford.

The second right rib, up near the top of the rib cage, curves around behind the clavicle, or collarbone. In the postmortem X-ray of Letha Rutherford's torso, the second rib was partially masked by the clavicle. That's why Dr. Wolf hadn't been able to see what I could now see and feel: a small, sharp nick in the rib, the sort of nick a thin-bladed knife might make as it slipped through the rib cage and into the lungs. I looked at it under a dissecting microscope, then showed it to Dr. Nichols and Murray, without commenting on what I had seen, and both of them reached the same conclusion I had: the location, direction of force, and shape of the notch in the bone appeared to have been made by a slicing or stabbing motion.

I'd seen numerous cut marks in human bones by this point in my career—scalped Native American skulls, surgically amputated limbs, crudely dismembered torsos. I'd perused dozens of micrographs of saw marks in human bones, made by one of my Ph.D. students, Steve Symes, in the course of a pioneering dissertation cataloguing the varieties of marks saw blades can leave as they push, pull, spin, and otherwise chew through the skeleton. This mark had clearly not been made by the back-and-forth motion of a handsaw, or the rotating blade of a circular saw; this cut mark had been inflicted in one motion with a sharp, thin-bladed implement.

I asked Dr. Nichols if I could take this rib back to UT for a closer look, and he agreed. He also let me take the hyoid so we could check for microscopic fractures that might tell us if Letha had been strangled. I signed evidence receipts for the two bones I thought might shed light on the girl's murder, and then we placed the rest of the bones in anatomical order in a new body bag. The following day—two days before Christmas—eighteen-year-old Letha Rutherford was buried for the second time. Mercifully, her waterlogged coffin was replaced by a new one, donated by a local funeral home.

Back in Knoxville, things were quiet at the university over the Christmas break, but several of my graduate assistants were around, taking advantage of the lull to catch up on research and writing. One of those was Emily Craig, a woman who'd worked for some years as a medical illustrator before deciding she wanted to become a forensic anthropologist and entering our Ph.D. program. For a closer look at the notched rib and the hyoid bone, I asked Emily to take the bones over to the Biology Department for a series of micrographs—extreme close-ups taken with an electron microscope, a powerful high-tech cousin of the optical microscope most of us have peered through at some point during our student years.

Credit for inventing the optical microscope, which bends rays of light through lenses to magnify objects, is often given to a Dutch textile merchant and scientist, Anton van Leeuwenhoek. Van Leeuwenhoek deserves tremendous credit, though not as the microscope's inventor. A Dutch spectacle maker named Hans Janssen claimed to have created the first compound (multiple-lens) microscope in 1590, though he didn't come forward with that claim until the mid 1600s. Decades before Janssen's claim, the Italian astronomer Galileo developed a compound microscope, by 1609 if not before. Van Leeuwenhoek began making "microscopes" in the mid-1600s, but his did not use multiple lenses; instead he developed a revolutionary technique for creating tiny, near-perfect spheres of glass which could magnify minuscule objects up to 500 ×. Although van Leeuwenhoek made neither the first nor the fanciest microscopes, he made better use of them than anyone before and perhaps—considering the technological limits he faced—anyone since. Over the course of some seven decades, van Leeuwenhoek examined things no human eye had ever seen before, including sperm cells and bacteria (a discovery the Royal Society refused to believe for years).

But as van Leeuwenhoek and his successors found, optical microscopes have limits. In theory, the wavelength of visible light limits an optical microscope's resolution or usable magnification to 1300 ×; in practice, though, the image turns fuzzy well before that, the way a grainy newspaper photograph turns into a meaningless pattern of big, blurry dots under high magnification. One way around this limitation is to look at specimens with electrons instead of with light. The electron microscope, invented in the 1930s, "illuminates" its target with a highly focused beam of electrons. Because electron beams are much finer than optical light, electron microscopes can magnify images millions of times, revealing structures as small as individual atoms! One type of electron microscope, the scanning electron microscope, or SEM, yields particularly crisp images of three-dimensional objects, at magnifications of 10,000× or more. (If you Google the words "scanning electron microscope SEM image gallery," you can visit web sites that show breathtaking details of insect antennae, spider claws, and other intricate microstructures.)

The SEM's resolution and clarity, I hoped, would reveal conclusively whether Letha Rutherford's hyoid had been fractured by a killer's grip . . . and whether the nick three forensic scientists had tentatively identified as a cut mark could have been a fatal stab wound. The Biology Department's scanning electron microscope had one other sophisticated capability. Besides providing remarkably vivid images of tiny three-dimensional features, the SEM could also be coupled with an element probe, an analytical instrument that can identify even minute quantities of chemicals present in a sample.

We'd used SEM micrographs in another case only a few weeks earlier, when the bodies of four women were found on a wooded hill beside 1-40 a few miles east of downtown Knoxville. In that case—a sensational series of murders for which a former Knoxville zoo employee, Thomas "Zoo Man" Huskey, was arrested—we found microscopic fractures in the hyoid of one of the four dead women. Given that one of Letha Rutherford's ribs suggested that she had been stabbed, I wasn't counting on finding evidence that she had been strangled as well, but over the years I've learned never to make assumptions.

Emily took the hyoid and the rib to Dr. David Joy, a microscopist in UT's Biology Department, who agreed to take SEM images and do X-ray element analysis of the rib's damaged region. The rib was first wrapped in parafilm to keep out contaminants. Then a short section of rib containing the nick was cut with a band saw and placed in a vacuum chamber; there it was coated with a film of carbon so thin as to be essentially invisible, to conduct electrical charges off the sample and keep them from interfering with the images. The hyoid, which was small enough to fit in the vacuum chamber in its entirety, also got a carbon coating.

The SEM images of the hyoid were completely normal. The bone itself was free of fractures; in addition, the cartilage that joins the projecting "horns" of the bone to the central arch, or body, appeared undamaged.

The rib was another story. The bone's outer layer, the cortex (also called "cortical bone") showed a thin, sharp, well-defined groove. It was clearly not caused by a dog's tooth: the long bones had plenty of bite marks, and they were nothing like this. It was also clearly not an antemortem fracture, as it showed no signs of healing. At a magnification of 6 ×, the cylindrical section of rib appeared nearly as thick as my wrist, with the cut slicing across roughly half of the rounded side of the bone that was visible in the image. At 14 ×, the sample looked as thick as my forearm, and the torn cortical bone fibers—all forced in the same direction, toward what would have been the upper lobe of Letha's right lung—looked like splinters in a freshly gashed tree branch.

Even those low-power images were potentially crucial evidence; they were completely consistent with a stab wound, and did not fit any other explanation I could think of. But those were features we could discern with a handheld magnifying glass. What we could never have discerned with a magnifying glass, or even with the world's best optical microscope, showed up when the cut was magnified thousands of times. Nestled within the groove in the bone was a pale, rounded particle that—magnified to 3,330 ×—looked to be about the size and shape of a piece of aquarium gravel.

In reality, it was less than one ten-thousandth of an inch in diameter. Another image, taken at nearly 5,000×, showed two additional, smaller particles, which appeared roughly the size and shape of rock salt. They were distinctly different from the fibrous bone, but what were they?

The answer came from the element probe, which used X-rays to identify the chemical signature of the particles—the specific wavelengths of energy the particles gave off when struck and excited by X-rays. The element analysis showed at least two metal particles not normally found in bone. One particle contained chromium, iron, and nickel—constituents of steel. The other particle contained cerium, a metallic element used to harden steel and also—in its oxide form—to make knife sharpeners (if you've ever seen a white, rod-shaped ceramic knife sharpener, chances are pretty good it was made of cerium oxide). The cortical bone adjoining the cut was checked carefully, but no additional, similar particles were found on the undamaged bone—only in the cut. The implications were chilling: the element analysis not only corroborated the evidence that Letha Rutherford had been stabbed to death, it also hinted that she might have been stabbed with a freshly sharpened knife.

On February 13, I sent a report detailing these findings to the Lexington police, the Kentucky M.E.'s office, and to Letha's mother, Brenda Wilson. Along with the written report, I included copies of the SEM micrographs and the X-ray element analysis. Detective Etherington, the lead investigator, forwarded this information to the FBI laboratory and asked if they could match the particles to a particular knife or knife sharpener. I heard a rumor, unofficially, that the police had their eye on a potential suspect, and that they had actually found a cerium knife sharpener in his kitchen, so I was hopeful that an arrest was imminent. But the FBI's response, which arrived two months after I submitted my report, was frustratingly noncommittal: "To determine the possible source of the cerium," wrote the Bureau's analyst, "both the distribution and form of the cerium should be determined. . . . Inasmuch as I have no information as to the crime scene or condition of the victim's body when found, environmental acquisition of a compound of cerium should not be overlooked." In other words, the analyst said, something from the trash heap where Letha's body was hidden might have contaminated the cut mark in her rib—without contaminating the surrounding, undamaged region of bone. As for the iron, chromium, and nickel, the FBI report read, "With the limited data presented resulting from presumably a bulk analysis, it cannot be assumed that all three constituents are combined or even related. . . . The data presented are not sufficient to support a conclusion that the three metals are in association, and if they are, that they are from a knife. In any event, a brand of knife would not be identified by any metallurgical analysis."

I wasn't ready to give up. Not long before, Knoxville's chief of police, Phil Keith, had consulted research scientists at Oak Ridge National Laboratory in a murder case. The Oak Ridge scientists had analyzed a smear of lead from the victim's body and successfully matched the lead's composition and trace contaminants—its chemical fingerprint, in other words—to one specific gun, which was owned by a suspect, even though they did not have traditional ballistics markings to compare. I relayed this information to the Lexington police and suggested that they consult the same scientists to see if they could perform a similar scientific feat in this case. As weeks turned to months, with no contact between the Lexington police and the researchers in Oak Ridge, my hopes for a breakthrough in the case began to fade.

In August 1993, eight months after we exhumed Letha Rutherford and found the cut mark containing traces of metal and cerium, I received a letter from Brenda Wilson. "I'm still hearing rumors," Letha's mother wrote, "that the second cousin who found Letha is making brags to other people that he's the one who killed her. I've been telling all this to [Detective] Etherington, but I haven't heard from him since you were up here."

I continued to hear from Mrs. Wilson from time to time over the next several years. She tried to interest several national television programs in airing a story about Letha's unsolved murder. I tried as well, but neither of us found any takers. Eventually I lost touch with her; I heard that she'd continued to struggle financially, and at some point when I dialed her telephone number, I got a recording saying the number had been disconnected.

In person and in her letters, Mrs. Wilson had repeatedly shared with me her belief that her daughter's murder simply fell through the cracks because the family was poor. I'm inclined to agree with her; if someone wealthy or prominent had been the victim, I suspect the authorities might have made more efforts to match the chemical fingerprint we found in the cut mark to a particular knife or sharpener. There's still no guarantee a match would have been possible. In fact, as a master's degree thesis project, a UT graduate student later tried using the SEM to detect the metallurgical "fingerprints" of knives in cut bones. She made dozens of cut marks in pig bones with several different knives, but she never found any trace elements left behind by the blades. Anytime you're pushing the capabilities of science and technology, her fruitless experiment reminded me, success is far from certain. Scientific proof requires reproducible results, and our results with the electron microscope weren't reproducible. Still, that doesn't mean they weren't correct.

Somewhere out there, perhaps just down the road from that dead-end dump and pot patch on Dry Branch Road, a murderer remains free. Somewhere in his house—maybe in a drawer of ordinary kitchen implements—the knife that killed Letha Rutherford might be tucked amid a jumble of other cutlery, lying atop a well-worn cerium sharpener. I don't know that, and I probably never will. But I'll probably never stop wondering about it. It's remarkable that science could bring us so close to solving a case that had gone so cold. But for Letha Rutherford and her mother, "close" just wasn't good enough.

12

LEOMA PATTERSON, PART 1: REOPENING AN OLD COFFIN—AND AN OLD CASE—WITH DNA

In 1953 British scientists James Watson and Francis Crick solved a puzzle that had baffled geneticists and chemists for decades: What was the structure of the huge molecule containing life's biochemical blueprints and operating manual? The answer, which Watson and Crick deduced from micrographs taken by Rosalind Franklin as she fired X-rays through crystallized specimens, was elegant but surprisingly simple. That molecule, deoxyribonucleic acid—commonly (and mercifully!) known simply as DNA—was shaped like a microscopic ladder three billion rungs high, its uprights corkscrewed into the double helix that is now one of the most familiar shapes on earth.

The breakthrough insight won Watson and Crick the Nobel Prize in 1962; it also paved the way, over the next three decades, for the complete mapping of all three billion rungs of the twisted ladder, and the development of analytical techniques capable of isolating any individual's unique genetic "fingerprint." But capability, in real-life forensics, doesn't necessarily lead to reality.

Redoak Mountain isn't particularly noteworthy, as mountains go. It tops out at 3,200 feet, less than half the height of some of the peaks in the Great Smoky Mountains, fifty miles to the southeast. But somehow, through some startling confluence of geology, bulldozing, and paving, the narrow switchback road I was on seemed to hang about 3,199 of those 3,200 feet above harrowingly steep, breathtakingly wild valleys. And "guardrail" seemed to be an idea whose time had not yet come to the southern end of the Cumberland Mountains.

Even the place names out here, a mere thirty miles northwest of Knoxville, harked back to an earlier, wilder world: Bearwallow Branch. Backbone Ridge. Graves Gap. One-room churches and tiny cemeteries seemed to outnumber the houses, unless you counted the handful of mobile homes, which appeared to have been airlifted into their notches in the steep hillsides. The route we were following over Redoak Mountain was so far off the beaten track, even the guy I was following—a local who supposedly knew where we were going—got lost, taking us ten miles in the wrong direction before reversing course and turning upslope.

The hardscrabble sprinkling of civilization clinging to the slopes and hollows of Redoak Mountain might logically raise the question of why?—why bother carving roads and snaking trailers into these remote backwoods areas? The answer came from the heavy trucks moaning down the grades. The Cumberlands might be demographically poor, but they're rich in minerals, and the trucks, laden with strip-mined coal for the generating plants of the Tennessee Valley Authority, were an intimidating reminder of the reason for most of the roads and perhaps many of the gravestones.

It was a muggy August morning in 2005, and I was on my way over Graves Gap and down the north side of Redoak Mountain to exhume a grave in a family cemetery in a remote, rugged corner of Anderson County. I'd gotten a call three months before from a recent UT graduate named Michelle Adkins, who had heard me guest-lecture in chemistry class one day. Michelle asked if I might be able to help answer a long-standing question for her family, one created by the disappearance and apparent murder of her grandmother more than twenty-five years before. She had my attention, and I told her I was willing to try. Michelle gave me some of the background, and a few days later one of her aunts, Frankie Davis, called from Texas to fill in more details. The story the two women told me was this: Leoma Patterson—Michelle's grandmother, and Frankie's mother—went missing from the town of Clinton, located about twenty miles north of Knoxville, back in October 1978. Five months later, some children playing along the bank of Norris Lake found part of a human skeleton. Animals had scattered or eaten most of the bones; what remained on the wooded hillside was a skull, some ribs, a couple of long bones, and a section of the spine and neck. The hair mat, which had sloughed off the skull as the scalp decayed, lay nearby as well.

Dr. Cleland Blake, a medical examiner who served several East Tennessee counties, examined the skeletal remains and tentatively identified them as Leoma Patterson's. It would have been difficult to make a positive identification in this case. There were no fingerprints to compare; for one thing, the corpse's hands were missing, and for another, Leoma Patterson had no fingerprints on file. She also had no dental or medical X-rays on file, nor any other unique skeletal features (healed fractures, for example) that her family knew of.

The case remained unsolved for more than five years; meanwhile, Dr. Blake held on to the bones, against the possibility that the case might take a turn or come to trial someday. And indeed, both of those came to pass: in 1985 a great-nephew of Leoma Patterson's, Jimmy Ray Maggard, confessed to killing her—they'd quarreled over a drug deal, he said, and the nineteen-year-old Maggard struck the fifty-six-year-old Patterson with a tire iron. The case was closed, Maggard pled no contest to voluntary manslaughter, and the skeletal remains were finally buried in September 1985, with a headstone bearing Leoma Patterson's name to mark the grave.

But Leoma's children—four daughters and three sons—were never completely convinced by the medical examiner's report. Over the next two decades, they'd often wonder whether those really were their mother's bones in that hillside grave. Eventually, as their own children—Leoma's grandchildren—grew up, they heard the story of their grandmother's death, including the doubts about the identification. And by the time granddaughter Michelle Adkins heard me lecture in chemistry class, DNA testing had become routine, and television shows like *CSI* had made millions of people conversant with phrases like "mitochondrial DNA." Hell, the top forensic scientist on *CSI* even had a body farm where he studied human decomposition, according to occasional snatches of dialogue. So this had to be a rigorously researched, incredibly accurate show, right?

Well, not exactly.

I'm often asked what I think about *CSI* and similar shows; I always answer that those shows have done a great job of educating people about the importance of carefully preserving and searching a crime scene for evidence. They've also gotten countless young people interested in careers in forensic science; if I had a dollar for every student or parent who has come up to me in the past few years and expressed an interest in a forensic career, I'd be a rich man.

But, I have to say, there are some enormous differences between television forensics and real-life forensics. For one thing, on television, they always solve the crime in an hour (actually, in forty minutes, if you subtract the commercials). The investigators almost never feel the need to ask questions; they already seem to know everything. And on those rare occasions when they do feel the need to ask a question, they make one quick phone call and get an immediate answer. Similarly, they get DNA results almost instantly. Ask a hundred real-life homicide investigators how quickly they get DNA results, and I guarantee you not a one of them will tell you it's a swift process; in real life, it takes weeks or even months, because certified DNA labs are terribly backlogged.

There's such a gap between how forensic science is portrayed on TV and how it is practiced in real life, and in real murder trials, that beleaguered police officers and prosecutors have given the gap a name—"the CSI effect"—and almost anytime one of them utters the phrase among colleagues, head-shaking and eye-rolling and muttering are sure to follow.

So it was with some reluctance that I negotiated the winding road up and over and down the north side of Redoak Mountain to disinter a mountainside grave, and that reluctance wasn't entirely on account of the corkscrew switchbacks and dizzying drops. This rural family, back in this hardscrabble area, had gotten the idea—an idea whose seed had been sown and watered to a great degree by television—that modern forensic science could now shed bright new light on the fate of Leoma Patterson, who had disappeared a quarter century before. This was not an affluent family by any means; in fact, I gathered that they had scrimped and saved for quite a while to pay for the exhumation, skeletal examination, and DNA testing. I hoped their hope was not misplaced, and that I could help resolve their nagging doubts once and for all.

Shortly after my guide led me down the last rampart of Redoak Mountain and onto a short, straight stretch of road running along a valley floor, we turned off the pavement onto a steep gravel track angling up another hillside. A quarter mile up, we pulled off and parked behind a handful of other pickup trucks and SUVs. Leoma Patterson's four daughters were already at the cemetery; so were three of her granddaughters, along with sundry husbands and boyfriends and possibly a grandson or so—I didn't get a good handle on who the various quiet males were, because the women, two generations of Leoma's descendants, seemed to be running the show.

They'd arranged to have a backhoe trucked in, and by the time I got there, the grave was already excavated and the lid of the coffin exposed, though it remained sealed. I shook hands for a few minutes, described what I was about to do, and then clambered down into the grave to open the coffin and see what lay within.

The metal lid was slightly dented and somewhat rusted, but it opened easily. Inside, the fabric liner was soggy, stained with mold, and coming loose from the lid. There wasn't much else inside the coffin—only a dozen or so bones, tucked into a couple of plastic bags. I'd brought two assistants with me: Kate Spradley, a Ph.D. candidate in anthropology who was skilled in the use of UT's ForDisc software, and Jon Jefferson, my coauthor, to take photographs and make a list of the bones in the coffin as I removed and named them, one by one. Kate had brought along a laptop computer and a three-dimensional digitizer, which she would use to take measurements of the skull there in the field; she would later plug these measurements into ForDisc and see how ForDisc classified the skull—as white, black, Asian, or some ambiguous mix of races. The computer and digitizer would be powered by a portable generator, which I'd brought along in the back of my pickup.

This was my first forensic case of 2005, so—in keeping with the numbering system I'd been using for decades—I assigned it the case number 05-1, although I also mentally equated that with Leoma Patterson. As I took each bone out of the coffin, I identified and described it to the dozen or so family members gathered around the edge of the open grave. They, like me, were most intrigued by the skull. The skull was small in size and smooth—"gracile"—in texture; it also lacked the heavy brow ridges and prominent muscle markings that characterize male skulls. The upper edges of the eye orbits were sharp, the mastoid processes (behind the ears) were small, and the occipital bone, at the base of the skull, lacked the protruding bump that is common in men. In short, the skull was clearly female. So far, so good.

To estimate age, I looked at several features of the skull, including the teeth and the cranial sutures, the joints where the bony plates that form the skull lock together. The third molars, or wisdom teeth, generally erupt around eighteen years of age. Although there were no lower third molars, both of the upper third molars had erupted, which meant she was probably at least eighteen. The cranial sutures would help me pin down her age more precisely. At birth, our cranial vault consists of seven separate bones; by age five, these all knit together at the cranial sutures, zigzagging joints that interlock. Around age thirty, the sutures generally begin to fill in with bone and smooth out; eventually they may become completely invisible, or obliterated. The sutures don't allow age estimates to be made with pinpoint accuracy by any means, but they can often narrow the range to within a decade or so, which can help investigators considerably: it's much easier to search records for missing females age forty to fifty than for females age twenty to eighty.

The first of our cranial sutures to close is the basilor suture, located at the base of the skull, between the occipital bone and the sphenoid; the basilor suture begins closing around age twenty. The skull I held in my hand had a fully closed basilor suture; that told me the woman was probably twenty-five or older. As I studied the sutures along the top and sides of the skull—the coronal, sagittal, and lambdoidal sutures—I saw that they were fairly well obliterated. The zigzag lines that appear so stark in skulls from twenty-five-year-olds were only faintly discernible in this skull. Mentally comparing these sutures to the thousands of others I'd studied over the prior half century, I estimated the woman's age to be at least forty. But I didn't think she could be too much older than that, because she lacked the signs of skeletal wear and tear that are typical of the elderly. She still had all of her teeth—quite an accomplishment for someone growing up in the 1930s and 1940s, especially in the hills of East Tennessee! She also lacked osteoarthritic lipping, the buildup of jagged ridges of bony material along the edges of vertebrae and other joint surfaces. As we age, we all tend to develop some degree of osteoarthritic lipping; it's sort of the skeleton's version of the mineral deposits that gradually clog a house's water lines. Osteoarthritic lipping is a major contributor to the aching joints that plague the elderly. Whenever I show slides of severe osteoarthritic lipping, someone always asks if there's any way to prevent it. "Of *course* there is!" I exclaim. "Die young!" Well, this woman hadn't died young, but she did seem to have died in middle age, rather than old age.

The skull's nasal opening was fairly narrow, with a well-defined sill at the base of the opening—features strongly characteristic of whites. Another white feature was the high-bridged nose. These features, together with the vertical ("orthognic") mouth structure, indicated that 01-05 was probably Caucasoid, or at least mostly so.

Two other features of the skull—actually, one feature that was present and one that was absent—jumped out at me. The teeth, while surprisingly numerous and sound (except for a few small, unfilled cavities), showed considerable wear on their occlusal surfaces—the edges that do the work of biting food and fingernails and pencils and wooden cigar tips. When I fitted the mandible into its normal position beneath the skull, I saw why: this woman had an edge-to-edge bite; that is, rather than having a slight overbite, as most white people do, her top and bottom teeth lined up and made contact all the way around her mouth. During decades of chewing, those occlusal surfaces had done more than just tear meat and grind corn; they'd also ground away at each other, wearing down the enamel a tiny bit each year. Native Americans tend to have edge-to-edge bites, so I asked the family members whether Leoma Patterson might have had any Indian ancestors. Yes, someone told me, she was supposedly part Cherokee. So the edge-to-edge bite and occlusal wear seemed to support the notion that this was indeed Leoma.

But the absent feature seemed to undercut that notion, because the absent feature was skull trauma: this skull was completely intact and undamaged. Yet if Jimmy Ray Maggard's story was true, Leoma Patterson's skull had been bashed with a tire iron, an implement that usually leaves a fracture so sharp and recognizable that it's considered a "signature fracture." (Other implements that leave signature fractures include clawhammers and—I know this because of a skull in our forensic collection—golf putters whose heads are rectangular in cross section.) On the other hand, family members told me that Jimmy Ray Maggard had a tendency to tell outlandish lies, so it was hard to know how seriously to take his account of hitting his great-aunt in the head with a tire iron.

Most of the woman's postcranial bones—the bones below the skull—were missing, and of the ones that were present, many had been chewed by carnivores. Only two long bones remained, the right humerus and the left femur, and neither of these was complete. The distal (elbow) end of the humerus had been chewed off, and both ends of the femur were gone—testament to how irresistible dogs find the marrow that lies within the ends of the long bones. Dr. Blake's crime scene report had indicated that an intact left humerus had been found at the scene, but his autopsy report did not list this bone in his inventory of skeletal elements. I hoped this inconsistency would be resolved by the presence of a left humerus in the coffin, but there was none. Lacking a complete long bone, I would be unable to calculate the woman's stature.

Fifteen vertebrae were strung on a rotting loop of twine in a sort of macabre necklace, created not as ghoulish jewelry but simply as a practical way to keep the vertebrae together. The first four cervical (neck) vertebrae were missing, and so was the twelfth thoracic (chest) vertebra, but eleven of the twelve thoracic vertebrae were present. That didn't surprise me, as the crime scene report had indicated that much of the upper torso remained, including much of the right rib cage, which was still covered with leathery skin at the time the remains were found near the lakeshore.

All twelve of the right ribs were present, either entirely or partially, though many had been chewed by carnivores. The left ribs did not fare so well: only the head and neck regions—stubs, in other words—remained of seven left ribs; the other five were missing altogether. Two adjacent right ribs—either ribs six and seven or ribs seven and eight (it's hard to distinguish between the middle ribs, especially if they're incomplete)—showed signs of trauma that was *not* caused by carnivores. At mid-shaft, both ribs exhibited a hinge fracture, suggesting that a blow to the right chest caused these two ribs to fold inward.

There wasn't much to be learned from the humerus or the shaft of the femur; aside from the carnivore damage, there was no sign of perimortem trauma. Actually, the left femur had been recovered from a nearby yard, not from the site where the body had decayed, so it was at least possible, if unlikely, that this femoral shaft came from another unfortunate person who'd been murdered or dumped alongside Norris Lake.

To answer the question that had haunted Leoma Patterson's family all these years—was this really her?—would require DNA samples, and I'd come equipped to take those. When exposed to the elements, to bacteria, and perhaps to the body's own decomposition processes and products, DNA gradually degrades, so our best bet for good DNA samples would be from within molars or within the mid-shaft of the long bones. Using a pair of pliers, I carefully extracted two teeth from the jaws: the right first molar from the mandible, and the left second molar from the maxilla. I placed each tooth in a sterile plastic sample vial, which I labeled and gave to Frankie Davis, Leoma Patterson's youngest daughter, who'd made arrangements with a DNA lab to analyze the teeth and compare their genetic material with that of Frankie and one of her sisters.

Extracting the teeth was easy. What was slightly harder was getting a DNA sample from the long bones.

I'd brought a Stryker autopsy saw, along with a 100-foot extension cord. We plugged the cord into the portable generator I'd brought, and with Jon holding the bones steady atop Leoma Patterson's granite headstone, I cut small cross sections from the shafts of the right humerus and the left femur. I sealed each of these in a sample vial and handed them to Frankie Davis as well.

While I was collecting the DNA samples, Kate Spradley had set to work taking skull measurements. It made for quite a study in contrasts to see Kate's laptop and 3-D digitizing probe—a sophisticated, self-contained forensic computer system—perched on the tailgate of a pickup parked in a clearing on this rugged, remote mountainside. The probe made quick work of the cranial measurements. Unlike calipers, which have to be carefully aligned on two points simultaneously, the probe can simply be touched to one point after another; somehow (don't ask me how) the computer plots each point in 3-D space relative to every other point touched by the probe, so the software can easily calculate, say, the distance between the eye orbits, or the width and height of the nasal opening, or the degree of prognathism in the mouth structure, and so on. Back when I was a graduate student working at the Smithsonian Institution, taking thousands of skull measurements every week, I could make three to four cranial measurements per minute—as long as I had someone with me to record the measurements as I called them out. Using the 3-D probe, Kate could take the same number of measurements in a matter of seconds—single-handedly and with greater precision, because the digitizer's sharp tip is easier to position than the ends of calipers, which can slide off the skull's domed surfaces, or prove difficult to fit into small openings.

Once Kate had finished the measurements and I'd answered many more questions from Leoma Patterson's descendants, I sealed the skull and other bones back in the plastic bags I'd found them in, placed them back in the coffin, and closed the lid. Someone reached down to help me out of the grave, and the backhoe operator fired up the machine and began filling in the hole. I turned my truck around, eased down the gravel road to the valley floor, and then snaked back up and over and down Redoak Mountain once more.

A few days later I wrote my report. In the "Summary" section, I concluded, "There is nothing in the skeletal material that I looked at that would not be expected in the skeleton of a fifty-year-old white female with American Indian ancestry."

Four months after I wrote that, Frankie Davis called me from Texas. She'd just gotten a report from Gen-Quest, the DNA laboratory that compared the samples

places, and ten-foot aluminum poles to poke around in the shallows of the lake. Although they searched for hundreds of hours, and divers went down anytime the hooks or the poles detected something unusual, they never found the plane. With sidescan sonar, we could scan miles of riverbed relatively quickly; with luck, the technology would reveal the wreckage under the water. The idea sounded promising, if we could lay our hands on a sidescan sonar.

Sidescan sonar works by transmitting high-frequency sound waves through the water on either side of a boat or a torpedo-like sonar "towfish," then detecting the echoes that bounce back. The echoes—specifically, tiny differences in the amount of time it takes them to bounce back, as well as subtle differences in the strength or intensity of the echoes—are interpreted by computer software. The software generates a picture of what's on either side of the boat or towfish as it moves through the water. If you've ever seen an ultrasound image of a baby in the womb, you've seen a medical cousin, you might say, of a sidescan sonar image.

Sidescan sonar was first harnessed in a high-profile search by undersea explorer Jacques Cousteau, who used it in 1975 to locate the *Britannic*, a sister ship to the *Titanic*, in 400 feet of water off the Greek island of Kea. In 1976, sonar helped pinpoint the wreck of the Great Lakes freighter *Edmund Fitzgerald* in Lake Superior. In 1985, marine geologist Robert Ballard and a team of U.S. and French colleagues towed a side-scan sonar array across hundreds of square miles of the North Atlantic's floor to seek the *Titanic* herself; after weeks of searching, though, it was low-light video cameras that succeeded in spotting the celebrity ship-wreck. In 1987, sidescan sonar helped find the wreck of the *Central America*, a side-wheel steamer that sank in a storm off the South Carolina coast in 1857, taking twenty-one tons of gold from California to the bottom of the sea.

During the three decades since Cousteau first demonstrated sidescan sonar's ability to find large shipwrecks, the technology had steadily improved, and systems became simpler and less expensive. But "less expensive" is a relative term. Jon and I did some checking around, and at first the prospects looked discouraging. The sonar systems used by professional marine salvage companies—the main market for the technology—normally cost anywhere from $25,000 to about $50,000. We certainly couldn't afford to buy one, and we couldn't find anyplace to rent one, either. Then we heard that the Blount County Sheriff's Office had just bought a sidescan sonar, to use in searching for drowning victims, stolen cars that had been dumped in the river, and anything else that might require a detailed underwater search.

The keeper of the sonar was Lieutenant Jeff Burchfield, the training officer for the sheriff's office, as well as the head of their team of swift-water divers. Burchfield, it turned out, also had a long-standing interest in the Elmer Reynolds case; when Jon and I talked to him about the possibility of using the sonar to search the river, Burchfield offered not only the sonar but the divers from the sheriff's office and the Blount County Rescue Squad, which shares equipment, divers, and training with the sheriff's office.

On a chilly day in the spring of 2005, Lieutenant Burchfield and several colleagues launched a Zodiac inflatable boat and a small skiff into a cove off the main channel of the Tennessee River, several miles downstream from the spot where Elmer Reynolds' skull had been found in 1974. Huddled down against a biting wind, we bounced up the river to the vicinity of the Knoxville Boat Club; there they submerged the towfish—a bright yellow torpedo-shaped device that bore an uncanny resemblance to a surface-to-air missile.

But after a frustrating and bone-chilling hour, we were forced to admit defeat: the system, which was brand new, kept chewing up the thermal paper on which it was supposed to print continuous images of the river bottom. We would have to try again another day—a warmer day, I hoped.

Weeks passed and the weather warmed, but the Blount County Sheriff's Office kept getting sidetracked with lawbreakers and other distractions from our sonar search. Jon and I repeatedly lamented the lack of a sonar system of our own. Then we learned something interesting: an Alabama company called Humminbird, well known among boaters for its depth finders and fish finders (based on simpler "downlooking" sonar), had just developed and begun marketing a small sidescan sonar system. The Humminbird sonar—which mounted directly to a boat's hull, rather than requiring a separate, submersible towfish—was designed for serious anglers, rather than for professional salvage companies. Instead of a printer that produced a continous spool of paper, the Humminbird system displayed video-like images of underwater objects on a high-definition plasma display; its electronics also included a built-in GPS to save the coordinates of any point of interest, plus a memory card that could save specific images. Best of all, the system was priced at about $1,700—less than one-tenth the price of a basic towfish system. It sounded user-friendly, sophisticated, and affordable.

As it turned out, it was even more affordable than $1,700: through some burst of inspired eloquence (or maybe simply annoying persistence), we managed to persuade Humminbird to loan us one of their sidescan sonars. Jon drove down to the Humminbird factory—located beside a lake in Eufaula, Alabama—to retrieve the sonar and learn how to use it.

In hopes of narrowing the search, Jon and Lieutenant Burchfield—who remained keenly interested in the case—arranged a meeting with a group of engineers and hydrologists at the Tennessee Valley Authority (TVA) to see if they could estimate where the airplane might be, given where the skull had washed up. The group's consensus was that the plane could be virtually anywhere. "To show you how hard it is to find an airplane in the river," said one engineer, "three years ago, a plane crashed in Watts Bar Lake just below Thief Neck Island"—about eighty-five miles downriver from our search area. "The two people in the plane got out, and they were picked up by some people in a boat. So we knew right where that plane went down. A salvage crew came up from Atlanta three days later, with a sidescan sonar, and searched for days. They never found a trace of it. Then we sent one of our crews out with a sonar, and they couldn't find it either." The demoralizing bottom line, as TVA saw it: "You're looking for a needle in a haystack."

That wasn't our only challenge. The airplane we were looking for was a Piper Super Cub, built in the 1940s. A two-passenger plane, it had a wingspan of 35.5 feet, which made it fairly small but not tiny. The problem wasn't its dimensions but its materials: the plane had a tubular steel framework, covered by painted fabric. The engine block and engine cowling were metal, and so were the propeller, landing gear, and wing struts. Still, even if the tubular framework were still intact—and after thirty years, that was a mighty big if—a Piper Super Cub would not reflect sonar signals nearly as well as an all-metal aircraft like a Cessna would. That meant, I suppose, we were looking for a needle-shaped sliver of rust in a huge, watery haystack.

Still, we were determined to look. Shortly after that, in the summer of 2005, we found ourselves slowly crisscrossing the waters of Fort Loudoun Lake, as methodically as farmers plowing a field. A friend of Jon's, a retired software engineer named Pierre Dufour, had gotten interested in our project, and Pierre agreed to loan us his pontoon boat for the search. Pierre and I took turns piloting. The boat's big flat deck and long, widely spaced pontoons made it stable and comfortable, which was good, since the search ended up consuming weeks.

In theory, the Humminbird could scan a stretch of riverbed up to 700 feet wide on each pass—350 feet to either side of the boat. In practice, however, we needed to search on a far narrower scale, because at the 700-foot scale, a plane with a 35-foot wingspan would appear as a tiny speck on the screen. We decided to set the range at 125 feet per side—250 feet per pass—which meant that for every mile of river we wanted to search, we'd have to zigzag back and forth across the river more than twenty times. At three miles an hour, each mile of river would take several hours—longer, if we saw something that merited a closer look or warranted multiple passes from different angles.

And we had many miles of river to search. Reynolds and his passenger had taken off from Skyranch Airport, located between river miles 637 and 638 (meaning the airstrip is 637.5 miles upriver from the mouth of the Tennessee). The skull was found four miles downstream from Skyranch. And one eyewitness report from the 1970s—discounted at the time because it came from someone who was at a party drinking the evening of the flight—suggested the airplane might have struck the water a mile or so upriver from Skyranch, meaning we'd need to search at least five miles of the Tennessee River.

Then there was the Little River, which flows into the Tennessee downriver from Skyranch, but upriver from where the skull was found. The mouth of the Little River is nearly half a mile wide where it joins the Tennessee; to complicate matters further, the final two miles of Little River lie directly south of Skyranch, separated from the airstrip by a low ridge or peninsula (where more than one aircraft has come to grief over the years). We couldn't rule out the possibility that the plane had crashed in Little River rather than the Tennessee, since the skull could have tumbled down either waterway to the confluence, and thence down the Tennessee for another mile before coming to rest in the mud.

So that brought us to seven miles of search area, which seemed daunting. Then someone reported spotting bones in the shoreline beneath Baptist Hospital—bones that proved to be a teenage girl's. Baptist Hospital is located in downtown Knoxville, between the Gay Street Bridge and the Henley Bridge. It's ten miles upriver from Skyranch . . . and two miles downriver from the Downtown Island Airport. Was it possible that Elmer Reynolds, who had a reputation as a daredevil pilot, tried to fly under the Gay Street Bridge to impress his passenger, and misjudged the distance? That seemed unlikely—surely at least one person on foot or in a car would have noticed and reported an airplane hitting the bridge. What seemed at least somewhat plausible, though, was the possibility that Reynolds had flown Linda over downtown, where the lights would have been at their most impressive, then ran out of gas. In trying to glide to the Downtown Island Airport, he could have ended up in the river just beyond the runway (where, likewise, more than one aircraft has come to grief—including one in the fall of 2006).

With all of these possible scenarios, our area of uncertainty had expanded to a total of fourteen miles of the Tennessee, plus two miles of Little River. There had to be a way to narrow it back down; we simply could not search that much river.

I took another look at Elmer's skull, and the more I looked, the more convinced I became that it could not have tumbled down fourteen miles of river, or even half that far: the skull was virtually undamaged, and although I didn't know of any experiments that attempted to study how far skulls could be carried by current, and how much damage they might sustain along the way, I couldn't imagine it had bumped along very far. We did spend one afternoon scanning the river between Baptist Hospital and Downtown Island Airport; we spotted a handful of oil drums and two sunken boats, but nothing that remotely resembled an airplane's wreckage.

We also did some wide-angle scanning of the Little River, where we saw debris from wrecked docks or boathouses, but nothing promising. We decided to concentrate on the five miles of water between the Knoxville Boat Club—which is on the Knox County shore directly opposite the skull's location—and the area where the tipsy eyewitness claimed she saw a plane hit the water. (In an effort to get more details, we managed to track down who she was, and where she was living—an elderly woman by now, she had long since moved from her riverfront house to a condominium. Her phone number was unlisted, and a visit to the condominium complex yielded the sad and frustrating news that our eyewitness was now suffering from a severe case of Alzheimer's disease. "Honey, she doesn't even know she is in this world," was the way the manager put it.)

So we were back to relying on the technology, and on the uncertain hope that the plane's wreckage wasn't consumed by rust or covered by silt. As the summer wore on, we spent days weaving back and forth across the river, our parallel tracks zigzagging up the GPS screen like stitches up a baseball's seams.

In the area where the skull had been found, the Blount County shore was steep and rocky—bluffs, in most places, except for the small, debris-choked inlet where a small stream entered the river and where the skull (and lots of debris) washed up. Within ten yards of these bluffs, the channel plunged to forty feet in depth. That posed a greater challenge for the Humminbird: in shallow water—ten to twenty feet—we could set the range narrow enough to show detailed images (the seats in a sunken fishing boat, for instance). At forty to fifty feet, though, we had to widen the sonar's swath, and that meant any airplane wreckage would appear smaller.

Despite my confidence that the plane must have sunk somewhere near the skull, the Humminbird showed nothing but blank river bottom and sunken tree trunks in the immediate vicinity. Nor anywhere within a mile downriver. Nor a mile upriver.

Just upstream from Skyranch, where the eyewitness's house looked down from a hilltop onto a half-mile stretch of river, the sonar screen showed two jumbles that looked as if they could be mangled wreckage. Jeff Burchfield had offered to provide technical divers from the Blount County Sheriff's Office and Rescue Squad, and he was as good as his word. One weekend, two boatloads of divers came out, along with the towfish sonar, which took a closer look at the debris underwater before the divers went down. In both cases, the debris turned out to be piles of brush.

We kept searching; meanwhile, Lieutenant Burchfield left town for six weeks at the FBI Academy in

Quantico, Virginia. One day, about half a mile below Skyranch, Jon spotted something bright on the sonar screen—generally an indication of something hard and dense, highly reflective of sonar's acoustic signals. After numerous passes from a variety of angles and at tighter and tighter ranges, he'd gathered images that looked remarkably like the remnants of an aircraft wing and fuselage from one angle. In another pair of images, a narrow cylindrical shape intersected the long, wing-shaped structure at a 90-degree angle. "This could be it," he said.

Jon e-mailed the images to Dave Betts and Mark Gibson, Humminbird's sonar engineer and marketing director. Both agreed that the images looked promising; they even pointed out shapes that corresponded with air-intake openings on the engine cowling of a Piper Super Cub. Jon also emailed the images to Lieutenant Burchfield at the FBI Academy, where Burchfield and others agreed that the images were exciting. "I think you've got it," said Burchfield. Shortly after returning from Quantico, he organized a dive team.

On a warm summer Saturday morning, a small flotilla of boats headed out from a TVA boat ramp on Little River and turned up the main channel of the Tennessee. In addition to Pierre's pontoon boat, our armada included the Blount County Rescue Squad's skiff, piloted by the squad's captain, Ken Shelton; and the inflatable Zodiac, which would divert other river traffic a safe distance from the divers.

Once we reached the site marked on the GPS screen and anchored the pontoon boat and the skiff, Burchfield wriggled into a scuba "dry suit"; so did Roger "Murph" Murphy, a Maryville High School teacher who volunteered in his spare time as a rescue squad diver, sonar operator, firefighter, and sheriff's office chaplain. They strapped on air tanks and slipped into Fort Loudoun Lake. Visibility in the murky waters was limited to a foot or so; the current, which was running several miles an hour, made it difficult to anchor the boats securely over the spot where the "target" had been seen, and harder still for the divers to figure out where they were in relation to the boats and the target. They ranged farther and farther from the boats, their safety lines carefully monitored by other divers on board. Two tugs from a diver would mean "pay out more rope"; three tugs would indicate, "I've found something." Four or more tugs in quick succession would signal an underwater emergency—an SOS.

For many long minutes, there were no tugs, just trails of bubbles in the water. Finally Burchfield came up, but Murph remained under, nearing the end of his safety line. Then he gave three tugs, and the suspense on board the boat soared. He remained under for a maddening length of time, then finally his head broke the water. He pulled off his mask and held up his two index fingers in the shape of a plus sign. "It's two great big ol' trees," Murph said, "laying crisscrossed like this, with a big rope and a bunch of other crap tangled around them." A groan of disappointment went up from both boats.

By now we were getting discouraged, and beginning to doubt the Humminbird: if its electronics or software could make a tree trunk appear flat and rectangular on the screen, like a wing, could we really rely on it to show us accurate images of what lay beneath the water? Jon decided to give it the acid test: he headed down to Watts Bar Lake, eighty-five miles downriver, and rented a fishing boat from a marina near the spot where the TVA hydrologists had told him a two-seat Cessna 152 had crash-landed in the river three years before. On a map they'd pinpointed the place the witnesses had said the plane sank; it was at river mile 550, near the mouth of a small cove, at a place where the river channel was about fifty feet deep—not an ideal depth for the Humminbird, but on the other hand, the target was an all-metal airplane, which should reflect sonar signals well; the search area was well defined; and the plane was reportedly relatively undamaged in the watery crash landing.

The story of the crash was interesting in itself. The Cessna 152 (often called a C-152) is sort of like the old Volkswagen Beetle of airplanes: small, slow, simple, and inexpensive (as airplanes go) to buy and to operate. That made them popular with flight schools; over the past thirty or forty years, more U.S. pilots have learned to fly in C-152s and their predecessors, C-150s, than perhaps all other types of aircraft combined. The C-152 that had disappeared in Watts Bar belonged to Middle Tennessee State University (MTSU), which offers an undergraduate degree in aviation. How the plane ended up in the lake is a classic story of a lesson plan gone horribly wrong.

The plane was based at MTSU, in Murfreesboro, about 175 miles west of Knoxville; on the morning of September 24, 2002, an MTSU flight instructor and a student with a private pilot's license had made a cross-country training flight from Murfreesboro to Knoxville. After refueling, they took off for their flight home. Along the way, the instructor decided to give the student an "engine out" exercise: a simulated engine failure, meant to sharpen a pilot's ability to select an emergency landing field and put the plane on the correct glide path for a "dead stick" landing. Rather than simply reducing the engine rpm to idle, which is the normal way to simulate an engine failure, this instructor got creative. Reaching down between the seats to the airplane cockpit's floor, he surreptitiously switched the fuel-selector valve to "Off," stopping the flow of gasoline through the fuel lines and causing the engine to sputter to a halt.

As the student pilot frantically scanned the lakeshore's wooded hillsides for a meadow or pasture, the drill seemed like a real emergency. And it soon became one: when the instructor attempted to restart the engine, it would not start. The water was a less dangerous place to land than the treetops, so they lined the plane up with the river channel, brought it down as slowly as possible, braced for a jolt, and opened their doors so they could not jam if the plane's frame bent upon impact. When the landing gear snagged on the water, the plane yanked to a sudden stop and began to fill with water. Both men got free of their harnesses and out of the plane before it went under, but now they faced another emergency. The student, Walter Harper, couldn't swim, and they were a quarter mile from shore in fifty feet of water. The flight instructor helped Harper stay afloat, but quickly grew exhausted from the effort. Harper urged him to let go and save himself. Luckily, just as the young man was sinking beneath the surface, a fishing boat swooped in from nearby. Strong hands hauled Harper and his instructor out of the water as their airplane disappeared beneath into the depths of Watts Bar Lake around river mile 550.

River mile 550 was where Jon began his search with the Humminbird sonar, even though the TVA staffers had assured him the plane was nowhere around. Just as in Fort Loudoun Lake, he crisscrossed the channel as if the rented fishing boat were a tractor or a lawn mower, working his way upstream toward Thief Neck Island, a mile away. He had one hand on the small outboard motor's tiller and the other on the controls of the sonar. By the time he reached the island's lower tip, a stiff wind was whipping the water into whitecaps, and storm clouds were gathering. Jon turned the boat downriver. Then, just as he reached mile 550 again, the wind seemed to lessen and the clouds lightened. Jon resumed his search pattern, this time working downstream from his original starting point.

Thirty minutes after starting the downstream grid, he noticed a small but very bright white line on the sonar screen, contrasting sharply with the dark background of the river bottom. An even smaller line paralleled the first one, separated by a tiny gap. One of the things sonar operators key in on is extreme contrast, as metals tend to reflect signals much more strongly than mud or wood; another thing they watch for is shapes not usually found in nature: perfect circles, 90-degree or 45-degree angles, parallel lines. Jon marked the spot on the sonar's GPS screen and circled back, zooming in for a closer look.

What he saw at closer range looked like a giant lawn dart embedded in the floor of the lake. Fifty feet down, he saw the unmistakable shape of a Cessna 152—its main wing, its tricycle landing gear, its entire fuselage, and the three vanes of its tail structure. He even saw a three-vaned acoustic "shadow," cast on the river bottom by the tail structure. On another pass, from a slightly different angle, he saw what appeared to be a furrow or gouge in the river bottom, ending at the front of the aircraft. When he compared the orientation of the image to the direction of the river's current, he realized he was seeing a trench that the plane's nose had dug in the mud as the current pushed it downstream a bit. From the black-and-white sidescan images, as well as a down-looking depth-finder image, he guessed that the plane was lying almost upside down, its nose and the leading edge of its high wing resting on the bottom at fifty feet, with the tail angling up about ten feet off the bottom.

After the disappointment of the crisscrossed trees, it would have been understandable had the people from Humminbird or the Blount County dive team been skeptical of another set of "airplane" images. And doubtless they studied them closely. But the details appeared too sharp and too specific to match anything but the features of a Cessna's underbelly and tail, especially in view of the fact that it was within a half mile of the reported sinking.

A few weeks later, another small flotilla laden with diving gear set out across a TVA lake. This time, there was no groan of disappointment, and just a few hours after a diver attached a line to the tail of the sunken aircraft, a set of inflatable lift bags tugged the plane free of the mud and hoisted it to the surface. Another twelve hours later, the plane—muddy but unbowed—was towed up a boat-launch ramp and parked in the grass to drain, and the next day an aircraft mechanic came to begin taking off the wings so he could transport the craft to his shop, where he planned to use its engine and left wing on a C-152 he was rebuilding.

We still hadn't found Elmer Reynolds' missing plane, but we'd proved that the Humminbird sonar—at less than one-tenth the cost of a commercial towfish sonar system—could be a powerful and cost-effective underwater search tool. And if Walter Harper and his instructor hadn't been lucky enough to get out of that Cessna when it hit the water, the sonar would have allowed us to bring some measure of closure to their families.

We had one other question about the Humminbird, and about sidescan sonar in general: we knew that more and more rescue squads and law enforcement agencies were looking to sonar for help in recovering drowning victims, but we found that no one had done a controlled scientific experiment to test sonar's ability to image the human body in water. We also knew that Humminbird was getting calls from cost-conscious search-and-rescue groups—which tend to be funded largely by donations—asking if their system could show a body underwater. Humminbird's answer was, "We think so, but we don't know."

We decided to find out if it could, and at the same time to see how it compared with a commercial towfish sonar. We also thought it would be interesting and helpful to see if the mannequins used by rescue squads as stand-ins for human drowning victims looked the same on sonar as a human body; that is, could sonar crews actually hone their skills using mannequins, or might the mannequins (which were made of various kinds of plastic) be transparent to the sonar signals, and therefore invisible?

To test all these things at once, we requested the help of one of my former students, Dr. Lee Jantz, who is now on the Anthropology Department faculty, overseeing the osteology lab and coordinating the donation program at the Body Farm. We asked Lee to loan us a cadaver, which we proposed to submerge in the river, near several mannequins; we asked the Knox County Rescue Squad and the Blount County Sheriff's Office and Blount County Rescue Squad to bring boats, towfish sonars, and mannequins. Lee took some persuading, understandably; her initial response to the proposed experiment was something along the lines of "You have *got* to be kidding." Eventually, though—once we explained how we could restrict public access to an isolated stretch of river, and how we could make certain we wouldn't lose the cadaver—she agreed, and even offered to bring along a graduate student, Rebecca Wilson, to assist. The Knox County and Blount County squads were enthusiastic about the chance to train with a real body, so both groups turned out in force with divers, boat crews, and sonar operators. In addition, the Knox County Rescue Squad brought a "Rescue Randy," a 165-pound mannequin used mainly in firefighter training, and the Blount County Rescue Squad brought a lighter-weight, department store mannequin they used to practice water searches. Simulaids—the company that manufactures "Rescue Randy" and other types of mannequins—loaned us a water-rescue mannequin, which is hollow and fills with water when submerged; the weight of the water, along with some built-in ballast, approximates the weight of an adult victim in the water. Last but far from least, the Anthropology Department provided, under the watchful eye of Dr. Jantz and graduate student Rebecca Wilson, the body of a recently deceased woman who weighed around 200 pounds.

Now, with two rescue squads, three sonar systems, and a dozen or more technical divers on hand, we were pretty confident we weren't going to lose track of Lee's cadaver; still, putting a body in an open stretch of river isn't something to be undertaken lightly or casually. We'd identified a quiet stretch of Little River, several miles upstream from its confluence with the Tennessee, with a combination of attributes that made it a good site: there were no houses along the shoreline in the area; there was a small Blount County park, complete with a boat ramp, that the sheriff's office agreed to close to the public for the day; and there was a long, straight stretch of water of consistent depth, where we could space the mannequins and the cadaver about a hundred yards apart, so the various boats and sonars could make a single pass and image all the targets in swift succession.

First the rescue squad teams anchored the three different mannequins in place; then they fitted a weighted dive vest on the cadaver to submerge the body on the bottom; as a precaution, they attached a safety line and an anchor. When Lee saw the respect with which the divers handled the body and the care with which they rigged the weights and safety line, her fears about the body accidentally ending up in somebody's boathouse (and on the TV news) eased considerably.

The results of the experiment were fascinating. The department store mannequin—made of a lightweight plastic foam and fiberglass—was virtually invisible to the Humminbird as well as to both towfish sonars, even though the sonar operators knew exactly where to look for it. So for sonar training, the Blount County Rescue Squad would clearly need to use something else as a target. The hollow water-rescue mannequin also had a negligible sonar signature, which wasn't surprising, since much of its volume was filled with water.

The 165-pound "Rescue Randy," while not designed for use in water, showed up pretty well on all three sonars. His feet were touching bottom, and he settled in a standing position in the water, which was readily apparent on the paper printouts and the plasma display.

The cadaver, as all three systems showed, lay on her back on the river bottom, her feet slightly apart. Knowing where she was made her easy to spot; if she were in an unknown location anywhere within a large search area (as drowning victims on the Tennessee River often are), I'd think it would be tough to find her. The human body, after all, is about 70 percent water, so it's not the brightest and best (not the "hardest," as the sonar operators say) target in the world. Now, a knight in armor who got thrown off his horse from a bridge over the river: he would be a *terrific* sonar target.

Still, even if sonar's not a magic tool for finding drowning victims, it's a huge step forward, as the headlines showed in November 2006, when law enforcement personnel in Utah found *six* bodies in a lake while using sidescan sonar to search for a missing couple. It can also be extremely useful in alerting blackwater divers to underwater hazards before they descend into tricky currents and murky depths.

As for Elmer Reynolds and Linda Hendrick: I haven't given up hope that we'll find the rusted wreckage of that airplane and recover their bones. Sonar is getting better all the time—Humminbird has just come out with a new, more powerful unit, and the Blount County and Knox County rescue squads are both interested in continuing the search as part of their regular training exercises.

But the honest truth is, in real-life forensic cases—unlike TV forensics—no matter how sophisticated the search technology or how skilled the people wielding it, some people will never be found. It's a painful fact of life, and of death. Elmer Reynolds' skull has been with me for more than thirty years now; his missing bones, and those of Linda Hendrick, are never far from my thoughts, but they remain tantalizingly out of reach. For now, at least. Maybe not forever.

POSTSCRIPT

In the spring of 2007, two noteworthy endnotes to the Elmer Reynolds case occurred. Unfortunately, neither of them is "We found the plane." The first was this: The Reynolds family, learning that I was still searching for the wreckage, asked for a DNA confirmation of the identification dentist Bob Greer and I had made in 1974. A comparison of DNA from a tooth in the skull with a DNA sample provided by one of Elmer's children confirmed that, yes, this really was Elmer. The second: After the DNA confirmed the identification, the family asked for—and received—the skull that had been in the UT forensic collection for more than thirty years. Elmer, or at least his skull, is gone, but not forgotten. I miss him already.

14

LEOMA PATTERSON, PART 2: PUTTING A FACE ON THE DEAD

To identify an unknown victim through DNA fingerprinting—or conventional fingerprinting, for that matter—you must compare two sets of fingerprints: the unknown person's, and those of someone you think might prove a match. In the case of the bones we had twice exhumed from beneath Leoma Patterson's headstone, we had no idea where to look for a potential match. No other middle-aged white women had gone missing in the late 1970s in East Tennessee and remained missing. We searched the database of the National Crime Information Center for a possible match, but found nothing promising there. That left us with one last long shot for nudging a fading memory to the surface of someone's consciousness: creating and publicizing a facial reconstruction—restoring a face, and maybe a recognizable identity, to this bare skull.

Down one level from the offices and classrooms of the Anthropology Department—down in the very basement of Neyland Stadium—is a large room with one wall of windows, two computers, a pair of desks, half a dozen or so long work tables, and hundreds of Native American skeletons from the Great Plains, unearthed back in the 1950s and 1960s, when dams on the Missouri River began creating vast lakes and flooding Indian villages dating back centuries. This room, the osteology laboratory—"the bone lab," everyone in the department calls it—is where the skulls and skeletons of the unknown and the murdered get measured, examined, and discussed on a daily basis by the Anthropology Department's forensic faculty and graduate students. On any given day—in fact, on this particular day—the work tables include bones from several forensic cases, including an unknown teenage girl whose bones were found just fifty yards from the city maintenance garage in a city a hundred miles from Knoxville. The bone lab has a single door which is kept locked during the rare occasions when no one's there working on a case or studying bones.

Beside a wall opposite the door, tucked into a nook between the work tables and the racks of boxed-up Indian skeletons, sits a smaller table normally decorated with one or two striking sculptures: the dead brought back to life, in a way, by a talented young forensic artist named Joanna Hughes. Joanna is, as far as I know, the only person in the United States with a degree in forensic art. She earned it at UT, creating the program herself by combining traditional art and sculpture classes with extensive study of anatomy and anthropology. Joanna's rare but valuable talent is to restore the faces of the unknown dead, working only from bare skulls, raw clay, extensive knowledge, and artist's intuition.

Clay facial reconstructions are among the last long-shot efforts to identify the unknown: in most cases, by the time Joanna starts layering clay on bone, years have passed since the final smile or tear or moan or plea. Not surprisingly, the odds are slim that an identification will ever result from an artist's best guess, in clay, at the face a skull once carried, years before. That makes Joanna's track record rather remarkable: of the nine reconstructions she's done in the past five years, three have led people who saw photographs of the sculptures to come forward and say, "I know who that is." Or, more precisely, "who that was."

The DNA report from GenQuest, which concluded that the DNA from the bones and teeth in Leoma Patterson's grave did not match the DNA from two of Leoma's daughters, opened a frustrating but fascinating can of worms. Instead of confirming that one old, closed case had indeed been solved, the DNA report raised two maddening questions: Where were the "real" Leoma's bones? And whose bones had we removed from her grave? With luck and a skillful reconstruction, Joanna might help us answer the second question.

Paul Phillips—the district attorney for Campbell County, where the bones had been found—opened a new case file and assigned a Tennessee Bureau of Investigation agent, Steve Vinsant, to the case. But whose case would Vinsant be investigating? Whose killer might Paul Phillips someday prosecute? To help answer those questions, I took the skull to Joanna in hopes that a reconstruction—and a photo in the newspaper or on the evening news—might trigger a thirty-year-old memory of a long-missing woman, and allow us to identify the woman who'd actually been buried in Leoma Patterson's grave.

It's often interesting to learn why someone gravitated to a forensic career, or has remained committed to it. Joanna Hughes's story is fascinating, heartrending, and ultimately inspiring. She grew up in a small town in south Alabama, Monroeville, population 6,690. Monroeville's main claim to fame thus far is that it's the hometown of the novelist Harper Lee, as well as the fictionalized setting of her classic novel, *To Kill a Mockingbird.* Monroeville was also the boyhood home of Lee's friend Truman Capote, the author of *In Cold Blood.* With its tree-lined streets, columned antebellum houses, elegant 1903 courthouse, and quaint general store ("O. B. Finklea Store No. 2, Dealing in What You Want—Dry Goods, Shoes, Groceries, School Supplies, Drugs, Hardware, Fertilizer"), Monroeville could, at least in places, easily pass for the 1930s town Harper Lee described as "an old town . . . a tired old town when I first knew it." By the time Joanna came along, it was even older and more tired.

Joanna never knew her biological parents. As a baby, she was adopted by Tim and Nancy Jones. Tim was the quintessential small-town doctor: he worked long hours, treated wealthy and poor patients alike (except that he didn't charge the poor ones as much), and held a place of both respect and affection. His wife Nancy was a one-woman civic institution: she made home-cooked meals for new residents, delivered toys to their children, and baked bread for old folks. A deeply religious woman, she trusted in the transforming power of faith.

At age twelve, Joanna startled her parents by announcing, "Someday I want to put faces on skulls." Looking back, she has no idea where she got such a notion; she only recalls that the vision was quite specific and crystal clear. What was less clear was how to turn it into reality—that is to say, into a paying job. So when it came time for college, she opted for something slightly more practical: a degree in film. At age nineteen, after her first year of film school at Florida State, she landed a job as a production assistant on the Steven Seagal film *Under Siege*. It was her first brush with Hollywood, and it was an eye-opening experience. She was the only young woman on the set, and she found herself receiving more attention, of a more personal nature, than she'd expected or wanted. Soon after completing her degree in film, she took a job as a secretary to earn a living, enrolled in art and anthropology classes at UT, and persuaded university officials to let her design a program custom-tailored to constitute a bachelor's degree in forensic art. In the fall of 2000, she began doing facial reconstructions on a volunteer basis in the Anthropology Department. The following year, in a National Geographic documentary about the Body Farm, Joanna put her art to the test before an audience of millions. Handed the bare skull of a man whose decomposition the documentary crew had filmed from start to finish, Joanna—knowing only that he was an elderly male of Greek ancestry (knowing, that is, only the sort of details police or a forensic anthropologist would be able to give her if the man had been an unknown murder victim)—re-created an astonishing clay likeness of the man's face. Opening a sealed envelope she'd been given once the reconstruction was finished, she pulled out a photo and saw what her subject had looked like in life. Her face beaming, she simply said, "It's *him*!"

Impressive as it was, that was just a demonstration. In the fall of 2002 Joanna got her first actual forensic cases: the skulls of two murdered African-American women, their decomposed bodies found together in the woods near Petersburg, Virginia, earlier that year. Two years after police and the Doe Network (a nonprofit organization dedicated to identifying unknown victims) published the photos of Joanna's reconstructions, the first of the two women was identified; in 2006, so was the second.

Joanna found putting faces on the dead to be intellectually and artistically satisfying, but not financially rewarding; each reconstruction took two weeks of concentrated effort, and she wasn't being paid. She began to consider going back to film production. Then, overnight, Joanna's life took a shattering turn.

Joanna wasn't the only child who'd been adopted by Tim and Nancy Jones. When she was a toddler, her parents adopted another baby, a boy they named Timothy Jason. Joanna took an instant dislike to Jason, her new brother. "I told them there was something wrong with him," she recalls. "I told them to take him back and get a different one." At some level, she'd sensed something frightening in Jason, and over the next two decades, her intuition would be borne out time and time again. Jason was an angry child, prone to violent outbursts—a source of bewilderment and sadness to parents who were kindhearted and gentle by nature. By his teens, Jason was using drugs and failing high school. He dropped out, though he later earned a GED (general equivalency diploma). He spent years drifting between short-lived jobs, drug rehab clinics, and occasional stints in jail. Joanna was a frequent caller to the Monroeville police; her calls tended to go something like this: "Hey, this is Joanna; could you come to my parents' house? Jason stole something again." But Tim and Nancy Jones refused to give up on him, and kept trying to help him turn his life around.

They kept trying until January 29, 2004. Early that morning, as Dr. Jones was leaving for his clinic, Jason bludgeoned his adoptive father to death in the driveway. Then he went inside and beat his mother to death as she lay in bed. Nancy Jones was bludgeoned so savagely, according to the *Mobile Press-Register*, that "parts of her jaw and teeth were found scattered across the room." After the murders, Jason took money from Nancy's purse, went out and bought crack cocaine, and returned to the house and smoked it. Then he set out in Nancy's car for a girlfriend's house in north Alabama. The car and the trip were apparently the motive for Jason's murderous rage: his parents had recently taken back an SUV they'd been letting him drive, and they refused to give him the keys for the trip, because that trip would have violated the terms of his probation.

At Jason Jones' murder trial in Birmingham, Joanna testified for the prosecution about her brother's chronic drug use and violent temper. She felt relief when he was convicted, and a grim sense of justice when he received the death sentence. And she felt icy satisfaction in September 2006, when Jason—on death row at Holman Prison in Atmore, Alabama—slashed his own neck with a makeshift blade and bled to death in his cell.

In a final twist of fate, Joanna—Jason's next of kin—was asked by the prison authorities what they should do with his body. Her answer, stemming partly from a desire to contribute to science and partly from an impulse for posthumous justice: Send him to the Body Farm. They did.

The tragedy in her own family has cemented Joanna's commitment to a forensic career. Although she can't bring back her parents, she can potentially help bring killers to justice, by restoring faces and identities to unknown murder victims—victims like the mysterious woman found alongside Norris Lake in 1979, a woman whom a DNA test concluded was *not* Leoma Patterson, despite her two decades spent beneath Leoma's headstone.

Joanna began the reconstruction by simply looking at the skull—noticing its general shape and unique features in something of the same way I key in on distinctive traits in individual bones. Two characteristics caught her trained eye right away: the high, wide bridge of the nose, and the broad, strong chin. Most women have sharp chins, but not this one.

Then came a tedious, time-consuming step in the process: cutting tissue-depth markers and gluing them onto the skull, to guide her as she applied the clay and began to shape its contours. To understand the importance of this step, do a simple experiment: Press the center of your forehead with a fingertip; there's not much tissue between your finger and the frontal bone, is there? Now feel your cheekbones: there's more flesh on those, but still not a lot. Now drop down between your cheekbones and lower jaw: there's a lot of flesh and muscle there, including the muscles you use to smile, and those you use to chew. Starting in the late 1800s, a series of German anatomists took careful measurements, from thousands of cadavers, of the thickness of tissue at numerous points on the head, especially the face. Their technique was simple but ingenious: they inserted a needle through a small cork, until the tip of the needle was just starting to emerge through the other side of the cork. Then they pushed the needle through the skin until it hit bone at each of dozens of landmarks on each cadaver's skull—landmarks with catchy names like "nasion" (where the nasal bone joins the skull), "pogonion" (the most anterior, or forward, point at the center of the chin), "bregma" (the top of the skull, where the frontal bone joins the two parietal bones), "ectomolare" (the point in the upper jaw where the second molar meets the bone), and "glabella" (the most forward point in the forehead, between the ridges above the eyes). They carefully withdrew the needle and measured how far the tip of the needle extended beyond the cork, which had stopped at the surface of the skin. Being Germans, the needle-wielding anatomists were meticulous and thorough, gathering data on men, women, and children of various ages and races. (However, their simple, ingenious measurement technique did have one built-in flaw: the slight bit of friction between the needle and the cork caused the cork to depress the skin slightly as the needle was inserted, so their measurements consistently understated the thickness of the tissue—a shortcoming that can now be avoided by measuring tissue thickness with magnetic resonance imaging: MRI studies.)

For her work, Joanna has the benefit of the original German studies, as well as modern MRI measurements. Using the data for adult women of European descent, she measured and cut depth markers from long, cylindrical erasers, the type used in Click-It mechanical erasers by artists and architects, then glued them to seventeen landmarks on the skull. Her markers for the white female skull originally identified as Leoma Patterson's ranged from a mere 2.75 millimeters (just over a tenth of an inch) at the end of the nasal bones to 17 millimeters (two-thirds of an inch) alongside the occlusal line, the fleshy region where the upper and lower teeth meet. Cutting the markers and gluing them onto the skull took hours, but without them, she'd be working at random, with no idea how closely the clay was approximating the depth of human tissue.

It's a painstaking process, one that seems a prime candidate for computerizing. You might think it would be a simple matter to program a computer to put a face on a skull—after all, isn't it simply a matter of adding so many millimeters here, so many there, smoothing the transitions, tinting the surface, and plugging in eyes, lips, and a nose? Well, it might not be terribly hard to program a computer to do those things, but it's devilishly difficult to program one to create a lifelike result. Over the years, I've looked at the results from several attempts to computerize facial reconstruction, and invariably they look like Claymation figures or lifeless masks. Science can provide the foundation for a good facial reconstruction, but—at least from what I'd seen so far over the years—only art can bring it to life.

I took the skull to Joanna in December 2006, shortly before Christmas. In mid-January, she was ready to show me what she'd done so far. It wasn't finished, she said, but it was getting there.

When I walked into the bone lab and saw what she'd done, you could have knocked me over with a feather. Joanna's reconstruction—based solely on the shape of the skull, published data on tissue depth, and the information I'd given her about the age and race of the woman exhumed from Leoma Patterson's grave—bore an uncanny resemblance to Leoma's daughter Barbara. It also looked strikingly like a photo of Leoma Patterson herself when she was in her twenties—before her face puffed up and sagged, as it had in photos taken in her forties and fifties. The more I compared the clay reconstruction with the photo of Leoma, the more remarkable the resemblance seemed—and the more troubling. What were the odds that two women, both in their fifties, would go missing in the same area of East Tennessee . . . and that they would look enough alike to be able to pass for sisters?

I'd promised to send pictures of the reconstruction to various members of the Patterson family as soon as they were available, and Joanna made it easy, by taking digital photos and emailing them to Jon, who is more computer-savvy than I am. Jon, in turn, relayed them to three of Leoma's children—Frankie Davis, Barbara Adkins, and Ronnie Patterson (who appears in the fifty-year-old photo as a small and very solemn-faced baby, sitting on Leoma's lap)—and to two of her grandchildren, J. R. Roach (the son of Leoma's oldest daughter) and Nancy Albert (the daughter of her youngest daughter). Nancy was the first to weigh in. "I was shocked," she said. "As soon as I saw it, I thought, 'That looks like my mother.' "

J.R.—at forty-four, the oldest of the grandchildren—was next. He was far less certain. "I don't see a whole lot of resemblance," he said.

Barbara was the daughter I thought the reconstruction most resembled—"dead ringer" (no pun intended) was the phrase that came to mind. So I was startled when Barbara said, "That's not her." Her brother Ronnie then got on the phone. "That don't look anything like our mama," he said.

The strong difference of opinion took me by surprise, though maybe it shouldn't have. Memory is a tricky, unreliable thing, facial reconstructions are approximations at best, and resemblance is in the eye of the beholder. What's more, there had always been some disagreement within the family about the original identification: Frankie Davis and her daughter Nancy were convinced that a turquoise ring found with the remains in 1979 had belonged to Leoma ("I saw a picture of my grandmother wearing that ring," Nancy says), but other descendants insisted that the only ring Leoma ever wore was one her brother had fashioned for her from a silver dime.

Within the Anthropology Department, opinion was more unanimous, though less passionate. Kate Spradley, the Ph.D. student who'd accompanied me on the first exhumation to take digital measurements of the skull, took one look and said, "That looks like the family members." And Lee Jantz, who runs the bone lab, compared the reconstruction to a photo of Leoma and raised her eyebrows, saying, "That is *very* interesting!"

But the DNA test had been conclusive. Or had it? Alarm bells began to ring in the back of my mind, and I took out the GenQuest report and pored over it. "Based on the above data," it read, "the profiles obtained for

Pearl Smith and Frankie Davis have identical sequence variations. It is likely that Pearl Smith and Frankie Davis are from the same maternal lineage." In other words, these two sisters had the same mother: that wasn't news to anyone. "In comparison to the Bone sample," the report went on, "Pearl Smith and Frankie Davis can be excluded as having the same maternal lineage." There, that part sounded conclusive, and exclusive: the woman whose teeth and bones I'd sampled was not the mother of Pearl and Frankie—was not, in other words, Leoma Patterson. In fact, the report sounded so conclusive we'd stopped referring to the dead woman as "Leoma" and had begun calling her "Not-Leoma."

But as I read and reread the report, other sentences began to take on more prominence in my mind. "The sample tested was a piece of bone." Why just a piece of bone, rather than either of the teeth? In a footnote to the data about the bone sample, I read this: "The presence of multiple probes prohibited the identification of a single mitochondrial DNA profile. Possible explanations include that the sample is degraded. It is not recommended to use this information for sole identification or comparison purposes." Finally, at the bottom of the report, was this note: "Direct sequencing of the HVI and HVII regions is recommended to determine conclusive results."

The more I studied the GenQuest report, the more contradictory it seemed. On the one hand, it seemed to say that the woman in the grave was not Leoma; on the other, it seemed to say that the conclusion wasn't particularly conclusive or reliable. DNA isn't my field of expertise—if you hand me a shattered skull or a knife-marked rib, I feel reasonably confident that I'll see what's forensically important there, but if you start talking about nucleotides or restriction enzymes, pretty soon it starts sounding like Greek to me. I decided to consult an expert, and luckily one was right at hand, unpacking her books and files.

At about the same time Joanna Hughes began gluing depth markers onto our baffling skull, Graciela Cabana, Ph.D., arrived in Knoxville and began unpacking boxes in her new office beneath Neyland Stadium. Graciela was an important addition to the Anthropology Department: a molecular anthropologist specializing in DNA. I desperately hoped Graciela could give me an objective reading and clear explanation of the GenQuest report.

We met in the office of Pam Poe, an administrative assistant in the Anthropology Department, because Graciela's office didn't yet have a desk, a phone, or even a chair. We were joined by Dr. Lee Jantz, who oversees the bone lab and the Body Farm's donation program, as well as Lee's husband Richard, who heads the Anthropology Department's Forensic Center. This case was a head-scratcher, and Lee and Richard were getting caught up in it just as I was.

Graciela began slowly, cautiously. "I can't quite tell," she said, "what GenQuest did and how they did it." She'd studied the report itself; what's more, she'd called the lab and—without indicating exactly why she was calling—had asked the lab's chief scientist some questions about their procedures and protocols. To be left wondering after talking to the senior scientist: that didn't sound good. "For one thing, I can't tell which sample they used. It appears they used only one bone sample, which surprises me. With old material like this, teeth are almost always a better source of DNA than bones." I'd known this—the tooth encapsulates the DNA and protects it from degradation better than bone—which is why I'd included two teeth in the set of four samples. As she elaborated on her questions about the GenQuest analysis, especially the inability to establish a single profile for the dead woman, Graciela's caution gradually dropped away. "This report is not just inconclusive," she finally said, "it's completely worthless." I was stunned. "The only thing you can tell from this report is the sample was either degraded or contaminated. Maybe both."

The air in Pam's office nearly crackled with electricity. Suddenly everything—every possibility—was up for grabs again. Originally the woman in the grave had been known as Leoma; after the GenQuest analysis, we'd taken to calling her Not-Leoma. Suddenly she'd become Maybe-Leoma. On television, DNA testing provides an instant, ironclad answer; in this real-life case, it had merely muddied the water: it appeared to provide an answer, but in fact, all it did was raise a whole new set of questions—not just for me, but for Leoma Patterson's family. They'd mistrusted the forensic system's original findings, which is why they'd contacted me in the first place; I couldn't imagine that this new twist would do much to restore their confidence. And it certainly wasn't likely to ease their pain.

Now what? One of Graciela's classmates in her Ph.D. program at the University of California at Davis was another molecular anthropologist, Jason Eshleman. After earning his doctorate, Jason had cofounded a company called Trace Genetics, which quickly established a reputation for extracting DNA from the most challenging samples possible: bones thousands of years old. "If anybody can get DNA for you," said Graciela, "it's Jason." I thanked her for the help, and vowed to contact him.

First, though, there was one other thing I wanted to try using the resources available to me at UT. It would require dismantling Joanna's beautiful sculpture, which seemed a shame—weeks of painstaking work, destroyed in an hour. But there was nothing more the reconstruction could tell us, and we had plenty of photos documenting the controversial face. Joanna cheerfully peeled off the clay, and once more I found myself confronted by a bare skull. It looked smaller and more exposed somehow, now that I'd seen a face on it. It also looked more cryptic, almost defiant. "Who am I?" she almost seemed to say. It was both a challenge and a taunt.

Joanna, Jon, and I—along with Maybe-Leoma—left the bone lab, climbed the long flight of steps flanking Neyland Stadium, and crossed Stadium Drive, then entered the basement of the Communications Building. Tom Owen, head of the university's television and photographic services, listened with interest as I described the twists and turns in the case, then agreed to help. He led us down the hall to UT's television studio, and his staff sprang into action.

We attached the skull to a stand that would allow us to raise, lower, tilt, and rotate it in any direction or plane. Then we set an ordinary music stand beside the skull and taped to it the best photo we had of Leoma—the one showing a young mother, firm of face, holding the solemn Ronnie on her lap. The studio crew rolled two studio television cameras into position, one zoomed in on the skull, the other on the photo. The images filled the screens of a bank of television monitors. Some showed just the skull, larger than life, some just the photo. But one monitor, the one I was most interested in, showed both images, one superimposed on the other—young Leoma Patterson's face half dissolved, the skull showing through the flesh, like some gruesome special-effects image in a horror film.

For over an hour we fiddled with the cameras, the photo, the skull: zooming in, zooming out, cocking the photo, rotating and angling the skull by fractions of millimeters. You know how portrait photographers always have you tuck your chin and tilt your head in a way that feels very unnatural and awkward? Well, the photographer who'd taken this photo of Leoma and Ronnie was no exception, and replicating that pose with the skull clamped to a wobbly Rube Goldberg–like assemblage of rods and knuckle clamps proved next to impossible. Finally, after dozens of adjustments and compromises—and with Joanna's hand wrapped in black to conceal it—we decided we had matched the images' sizes and angles as closely as humanly possible. The angles weren't a perfect match—we'd have needed a real, live, or at least three-dimensional Leoma to achieve that, and we didn't have her. That meant that the studio television camera couldn't reproduce the focal length and aspect ratio of the still photo with 100 percent accuracy. But it was darned close, and would certainly show us if the skull's main landmarks and basic shape seemed to line up with Leoma's . . . or if the superimposition revealed some obvious, exclusionary difference that would say "I'm not her": cheekbones that extended beyond the margins of the face, for instance, or a chin that was much too flat, or eye orbits that hovered up around the brow ridge.

I asked the technician in the control booth to show only Leoma's photo, then—very slowly—to dissolve to the skull. As the bone began to show through, I had him stop, and I felt the same tingle of excitement I'd felt when I first saw the facial reconstruction. Showing through the flesh of Leoma Patterson's strong, rounded chin was the center of the skull's strong, rounded chin—the landmark named "pogonian." Four inches above it, right between Leoma's eyes, the skull's nasal bone ran directly beneath Leoma Patterson's nose, with the landmark known as nasion positioned right where Leoma's would be. The correspondence between cheekbones and jawline was remarkably good, too.

Just weeks before, a DNA lab's report had me completely convinced me that the skull was not Leoma Patterson's. Now, I felt myself doing another U-turn.

"I think it's her," I said. Half a dozen heads in the television studio nodded in agreement.

The superimposition was persuasive. But it wasn't proof.

We tried another, high-tech route as well. As a master's thesis project, one of the Anthropology Department's graduate students, Diana Moyers, was working as a visiting scientist at the FBI laboratory in Quantico. Diana's project was to test and help refine an experimental computer program called "ReFace"—a facial-reconstruction system whose full name is "Reality Enhanced Facial Approximation by Computational Estimation." You can see why they shortened that to "ReFace." Developed for the FBI by GE Global Research, ReFace starts with skulls—three-dimensional CT scans of skulls, to be precise—and uses tissue-depth data and algorithms (fancy mathematical formulas) to do virtually, on a computer screen, what Joanna Hughes does with clay on a skull. Being a computer rather than an artist, ReFace works more objectively than Joanna does; it also works far faster, completing a 3-D rendering in a matter of hours rather than weeks. With its speed, ReFace has great potential for helping identify victims found in mass graves, as in Iraq. If it had been available at the time, it also could have helped enormously in the wake of Hurricane Katrina, when forensic teams struggled for months to identify hundreds of badly decomposed bodies. If computerized facial reconstructions of those storm victims could have been easily produced and widely distributed, the task of restoring names to the dead might have moved forward far faster.

Jon and I put in a plea to Diana and her FBI supervisor, Phil Williams. To our delight, they agreed to run Maybe-Leoma through ReFace; all they needed was a CT scan of the skull. That was easy to arrange, since one of the Anthropology graduate students, Megan Moore, had recently scanned every skeleton—more than six hundred—in the donated skeletal collection. On April 5, 2007, Megan and Todd Malone, a CT technician in the Radiology Department at UT Medical Center, ran the skull through the scanner, faceup in a box packed with foam peanuts. Megan FedExed the scans to Quantico, where Diana and Phil Williams ran them through the experimental software.

It was with high hopes, a few weeks later, that I studied the computer screen as it flashed up the likeness ReFace had overlaid, with mathematical precision, on top of the CT scan of Maybe-Leoma's skull. Surely this image, I thought—the fruit of several years of collaboration work by computer scientists, forensic artists, and anthropologists—would clearly answer the question: Was this Leoma, or was it Not-Leoma? Instead it merely amplified the question. The flesh-toned image on the computer screen—eyes closed, devoid of expression—could have been a department store mannequin, or a sphinx. There was nothing in the image, no matter how I rotated it in three dimensions, that said, "I am Leoma." Nor anything that said, "I am not Leoma." To paraphrase Winston Churchill's famous comment about Russia, the masklike face on the screen was "a riddle wrapped in a mystery inside an enigma." Between the scan, the software, and the tissue-depth data the software merged with the scan, it was all very sophisticated and high-tech. But it was still a damn mystery, and she was as enigmatic and elusive as ever.

Our only hope for proof was to go back to the DNA. We'd have to try climbing the twisted ladder one more time. And by now I realized just how twisted that ladder could be.

15

THE DAY THE BOPPER DIED: DIGGING FOR THE TRUTH, AND FINDING IT WITH X-RAYS

When your research lab is a body farm and your chosen field is the field of the dead, mysteries take many forms. So do techniques for solving them. And sometimes the techniques and solutions—sometimes even the mysteries themselves—don't present themselves until long after someone has died. Several of the cases in this book have required exhuming coffins and examining the skeletonized remains buried inside them: the weathered bones of Liz Wilson, Letha Rutherford, and the woman who might or might not be Leoma Patterson—people long since dead by the time they were found. But one recent effort to solve a mystery required digging up and analyzing extensive skeletal trauma in an embalmed corpse that had been buried just days after death—a famous, violent death that occurred nearly fifty years ago. I didn't know how the passage of so much time would have affected the body to be exhumed, but I suspected that the years had not been kind. I've exhumed too many decomposed bodies over the years to have much faith in embalming, at least as it's usually practiced.

There are ways to make a body last for decades or even centuries. Thousands of years ago, Egyptians learned that mummifying bodies—removing the internal organs and then drying the corpse rapidly—would sidestep the normal decomposition process. Two twentieth-century corpses designed for long-term display—Vladimir Lenin and Eva Perón—underwent elaborate embalming processes; to this day Lenin, I'm told, is topped off with embalming fluid during regularly scheduled maintenance. During the past several years, the remarkably posed corpses in two sensational cadaver exhibits, "Body Worlds" and "Bodies," have attracted long lines of viewers in every city they've visited. These bodies are preserved by plastination, a process that replaces all the water in the body's cells with silicone, transforming corruptible flesh into rugged plastic. I learned recently that some 7,500 people have signed up to donate their bodies to "Body Worlds" for plastination and display, a figure that makes the Body Farm's one thousand donors seem paltry by comparison. Given that these plastic cadavers could have nearly the same longevity as nuclear waste, I can't help but wonder two things: Where will they all end up? And what will archaeologists (or alien invaders) think when they unearth them thousands of years from now? I wasn't expecting such durability or plasticized perfection of the battered body that would be hoisted from the soggy coastal plain of Beaumont, Texas. . . .

One afternoon in the summer of 2006, when I came home from an errand, my wife Carol met me at the door, almost beside herself with excitement. "You just missed a call from the son of the Big Bopper," she announced.

"The son of the *who*?"

"The Big Bopper!"

I looked at her blankly and shrugged.

"You don't mean to tell me you don't know who the Big Bopper was?" I admitted that I didn't, and I could almost see my stock plummeting in Carol's eyes. "He's the singer who did 'Chantilly Lace,'" she said, and began to sing. "Chantilly lace and a pretty face . . ." Now I nodded, relieved; I did, at least, recognize the song. "He died in that plane crash with Buddy Holly and Ritchie Valens back in 1959."

Ah, 1959. That explained why I'd initially drawn such a blank. From 1956 to 1960, popular music passed me by completely, because I was living and breathing anthropology eighteen hours a day, seven days a week. I was racing to finish my Ph.D. at the University of Pennsylvania then, and my life revolved around teaching anatomy and working on my dissertation, except when I was catching a few hours of sleep. Pop culture, the beginnings of rock and roll, and teen idols like Holly, Valens, and the Bopper were simply not on my radar. At the time, I never heard Buddy Holly sing "Peggy Sue" or "That'll Be the Day"; I never tapped my feet to Ritchie Valens belting out "La Bamba"; I never sang along when the Big Bopper crooned "*Hel*-lo, *ba*-by" and launched into "Chantilly Lace." I was not aware, at the time, that a chartered Beechcraft Bonanza crashed into a frozen cornfield outside Clear Lake, Iowa, killing the pilot and three teen idols, Holly, Valens, and J. P. Richardson Jr., better known as "the Big Bopper." And I certainly could never have guessed that on a warm March morning forty-eight years later—in an effort to answer two haunting questions—I would find myself face to face with the mortal remains of the Big Bopper himself, hoisted from the soggy ground of a cemetery in Beaumont, Texas.

Beaumont sprawls across a few square miles of scrubby coastal plain some ninety miles northeast of Houston and twenty miles up the little Neches River from the Gulf of Mexico. Beaumont would be utterly unremarkable if not for two claims to fame. One of those is the Big Bopper. The other is Spindletop.

On the surface, Beaumont is flat and—geographically, at least—about as interesting as a parking lot. The first settlers of record, in 1824, were named Noah and Nancy Tevis, and their settlement was called Tevis Bluff. The name must have been either a joke or a reference to a gambling strategy, because the elevation in Beaumont ranges from a few inches above sea level to a whopping twenty-two feet. Texans brag about how big everything out there is, but I have to say, no self-respecting Tennessean would even *think* of using the word "bluff" to describe a landscape you could scarcely stub your toe on.

What's remarkable about Beaumont's landscape, aside from how utterly boring it is, isn't aboveground; it's below—a thousand feet below, and a hundred million or so years old. It came to light, in spectacular fashion, in January of 1901.

By the late 1800s, America's industrial engine was beginning to burn oil instead of coal, and that engine was quickly developing a powerful thirst. John D. Rockefeller became the world's richest man by catering to that thirst. But Rockefeller and other oil barons realized that petroleum reserves east of the Mississippi were limited, so wildcatters—armed with maps and reports from the U.S. Geological Survey (USGS)—began prospecting in the West. No one at the USGS thought Beaumont had anything to offer in the way of oil. But a Beaumont businessman named Patillo Higgins thought differently.

In the 1890s, Higgins—a one-armed logger, realtor, Sunday school teacher, and brick manufacturer (now *there's* a guy with an interesting résumé)—became convinced that an underground salt dome a few miles south of Beaumont held a vast pool of oil. His belief was based, at least in part, on an occasional trickle of oil from the ground there. Higgins and a series of investors spent most of a decade and most of their money drilling wells and coming up dry. Then, on January 10, 1901—just as the investors were running out of money and the drilling crew was running out of pipe—the ground at Spindletop began rumbling and the drilling platform started vibrating. To borrow from a song that wouldn't be written for another half century, there was a whole lot of shakin' goin' on. Then, with a deafening roar, six tons of drilling rig rocketed skyward, launched by a geyser of salt water, mud, gas, rock, and oil. Lots and lots of oil: 100,000 barrels (4,200,000 gallons) a day. It took nine days to cap the well, and during that time, a lake of Texas crude formed around the platform. Trainloads of tourists came to see the gusher; one enterprising local businessman sold them small bottles of oil from the lake at a dollar apiece. By the end of 1901, hundreds more wells—and many more gushers—had sprung up around Spindletop. Beaumont became such a boom town that Houston began touting itself as "the gateway to Beaumont." Gulf Oil, Texaco, and even Exxon Mobil can trace some of their corporate DNA to Spindletop and Beaumont. A second Spindletop strike—substantial, though not as sensational as the original gusher—occurred in 1925. But gradually the salt dome's production tapered off, and by the 1950s, Beaumont had gotten mighty quiet.

Then, in the spring of 1957, things got lively again. This time, the energy came not from oil, but from radio waves and charisma. A local radio announcer and disc jockey at WTRM-AM named Jiles Perry Richardson Jr.—he called himself "Jape" on his midday show, "The Dishwashers' Serenade"—started a new afternoon broadcast aimed at teenagers. Capitalizing on the craze for a dance called the Bop—a cross between the jitterbug, swing dancing, and the twist—Richardson transformed himself into a whole new character, a boisterous hipster he called "the Big Bopper," then undertook a big publicity stunt to prove he was worthy of the

Texas-sized name. One afternoon that May, he began a live broadcast on WTRM, and he didn't stop until five days, two hours, and eight minutes later. Taking only five-minute breaks during newscasts, the Big Bopper played 1,821 songs, set a world record for the longest live broadcast, and garnered nationwide news coverage. He also made connections that would help launch him as a songwriter and performer on the national stage.

Richardson formed a small record company and released a 45-rpm single later that year. The A side was titled "Beggar to a King," and the B side was called "Crazy Blues."

Neither song got much airplay beyond WTRM, but the Big Bopper was just getting warmed up. In June of 1958, he released a second 45; this one's A side was a forgettable song called "The Purple People Eater Meets the Witch Doctor," another effort that quickly faded. But the B side was another story altogether: the B side got so much airplay that the rights were snapped up by Mercury Records, one of the era's major labels, which put all its marketing muscle behind the song. Within weeks the song was at number 6 on the pop charts, and millions of American kids were bobbing and bouncing as the Big Bopper called "*Hel*-lo, *ba*-by! This is the Big Bopper," and then proceeded to sing the praises of teenage girls:

Chantilly lace and a pretty face
And a pony tail hanging down
That wiggle in the walk and giggle in the talk
Makes the world go round.

"Chantilly Lace" was the third-most-played song of 1958; an early television video—the world's first music video, according to rock-and-roll historian Bill Griggs—shows why. In it, the Bopper hams it up big-time for the camera, looking alternately baffled and delighted as he listens, protests ("but . . . but . . . but, *ba*-by!"), and sings to the unseen female on the other end of the telephone line. Richardson had big plans: he wanted to own his own radio station, and form his own company to produce and distribute other music videos. First, though—to earn money for the ventures—the Big Bopper signed on for the Winter Dance Party Tour, a road trip whose other acts included headliner Buddy Holly, Latino sensation Ritchie Valens ("La Bamba"), Frankie Sardo, and Dion and the Belmonts.

Despite its festive name, the Winter Dance Party Tour was sheer torture, at least for the performers. They were traveling by bus—not the lavish motor homes used by bands today, but a dilapidated old school bus. During their first ten days on the road they performed every night and traveled more than

2,500 miles, crisscrossing the frigid, windswept northern Great Plains. One night the bus broke down on a deserted rural highway. Outside the bus, the temperature was 35 degrees below zero. Inside, it wasn't much warmer. Desperate to keep warm, some of the performers began burning newspaper in the aisle of the bus, creating an interesting dilemma: Which was more likely to kill them, hypothermia or asphyxiation? The whole group might well have died if two hardy performers hadn't hitched a ride to the nearest town and brought sheriff's deputies to the rescue. By then, drummer Carl Bunch had frostbite on his feet—he was hospitalized and dropped out of the tour at that point—and Ritchie Valens and the Big Bopper caught severe colds.

Fed up with the brutal bus travel, Buddy Holly chartered a small plane to fly him and two of his band members from Clear Lake, Iowa, to their next stop, in Moorhead, Minnesota (just outside Fargo, North Dakota). By aging bus, the 360-mile trip would take all night; the sleek, V-tailed Beechcraft Bonanza—registration number 3794N—would be able to make the trip in just two hours.

The Big Bopper, who was still sick, asked Holly's bass player—Waylon Jennings—if he could have Jennings' seat on the plane, and Jennings agreed.

Ritchie Valens, who'd never flown on a small plane before, pestered guitarist Tommy Allsup to give up his seat. Finally Allsup agreed to flip a coin to see which of them would get the last spot on the plane. Allsup flipped, and Valens called "heads." He was right—and fatally wrong.

The three musicians were ferried to the airport by the manager of Clear Lake's Surf Ballroom, where they had just performed for some two thousand fans. The plane, owned by Jerry Dwyer, would be piloted by twenty-one-year-old Roger Peterson. Peterson was considered a skilled pilot, although he was not certified to fly by instruments alone—and had actually failed an instrument check-ride a few months before. But Peterson had no reason to think he would encounter any weather bad enough to make instrument-flying necessary. What he didn't realize was that a cold front, clouds, and snow were closing in on Clear Lake and Fargo faster than previously forecast. At 11:30 P.M., the sky over the airport was clear below 6,000 feet; by midnight, the ceiling had dropped to 5,000 feet; and by the time the plane taxied into position for takeoff at 12:55 A.M. on February 3, the clouds had dropped to 3,000 feet, the wind was gusting to 35 miles an hour, and light snow was beginning to fall.

Peterson gunned the throttle and the plane tore into the wind. Jerry Dwyer watched his plane climb off the southbound runway, do a U-turn to the north, then—just beyond the airport—bank left, toward the northwest and Fargo. The plane's red tail beacon was clearly visible. A few miles from the airport, though, Dwyer saw the plane enter a low cloud; when it emerged moments later, it appeared to be descending. Dwyer hoped that the light was simply vanishing over the earth's curved horizon.

Roger Peterson had promised to radio the control tower and file a flight plan after takeoff, but as the minutes ticked by, there was nothing but silence. Silence, and the buffeting of the wind. At Dwyer's request, the air-traffic controller repeatedly hailed the Bonanza, but he got no answer.

At dawn, fearing the worst, Dwyer took off in another plane, traced the route he'd seen the Bonanza take, and spotted the wreckage in a field belonging to Albert Juhl. The aircraft had struck the frozen ground, skipped, and slid 540 feet across the snow-covered field, coming to rest against a barbed-wire fence. Pilot Roger Peterson's body had remained in the cockpit; the bodies of Holly and Valens were thrown clear but still close to the wreckage, while that of the Big Bopper lay farther away, on the other side of the fence.

News of the deaths stunned the youth of America (including young newspaper carrier Don McLean, who left the bad news on his customers' doorsteps—and immortalized the crash a decade later in the song "American Pie," where he coined the phrase "the day the music died"). It also stunned the families of the musicians, who first learned of the accident from the media—a particularly painful way to hear of the tragedy. In the wake of the accident, and the families' anger at not being contacted first by officials, the government adopted a policy of not releasing crash victims' names until after their families had been notified in private.

The Civil Aeronautics Board—the predecessor to the FAA and its National Transportation Safety Board—sent a team of investigators to Clear Lake; they combed the wreckage, reviewed weather data, and dug into the records of both the plane and the pilot. From the damage to the plane and the track it left across the snowy field, they determined that the right wingtip had hit the ground first, because the plane had been descending fast and steeply banked to the right.

The shattered flight instruments, their needles frozen at the moment of impact, recorded the plane's final instant. The tachometer read 2,200 rpm—a normal engine speed for cruising, several hundred rpm below full throttle. The adjustable-pitch propeller had sheared off at impact, but the central hub—which controlled the propeller's pitch—was likewise set for cruising (rather than for climbing, which uses a higher-thrust setting, like a lower gear on a car). The airspeed indicator was stuck between 165 and 170 mph, also normal for a Bonanza when cruising. But the rate-of-climb indicator, which tells how rapidly the plane is gaining or losing altitude, showed the aircraft dropping at 3,000 feet per minute.

Most telling of all, in the eyes of the investigators, the plane's attitude indicator, or artificial horizon, was stuck in a position that indicated the plane was in a 90-degree bank, with its right wing angled directly down at the ground and its left wing pointing straight up at the clouds overhead. At that attitude, the plane's wings would generate no upward lift at all; the engine and prop would continue to pull the plane forward at a normal cruise speed, even as the plane sliced edgewise through the air, plummeting to the ground.

But how could this have happened? How could an experienced pilot not have known that his aircraft was banked 90 degrees, its wings vertical rather than horizontal? Here's where things get interesting, subjective, and contentious. The crash report faulted Roger Peterson's judgment for flying into deteriorating weather. The technical language used, with appalling frequency in small-plane crash reports, is "visual flight into instrument meteorological conditions," and it's the number two cause of single-engine aircraft fatalities (second only to running out of gas). When a pilot can't see the horizon, it's virtually impossible to tell, without instruments, if the plane's flying straight and level; in fact, it can take only a matter of seconds to become fatally disoriented. (That's what seems to have happened to John F. Kennedy Jr.: flying to Martha's Vineyard on a hazy night, he lost sight of the horizon, got disoriented, and spiraled into the sea.)

There was one more strike against Roger Peterson during that nighttime takeoff into deteriorating weather, the crash report concluded: the attitude indicator, or artificial horizon, in Bonanza 3794N was not the kind most planes are equipped with, and not the kind Peterson knew best. I'm not a pilot, but my coauthor, Jon Jefferson, is, and Jon has explained artificial horizons to me to the point where my eyes began glazing over. Basically, an artificial horizon is an instrument whose dial shows a miniature plane flying toward a miniature horizon, where earth meets sky, and mimics the plane's orientation—its attitude—as the pilot maneuvers the aircraft. If the pilot banks to the right, the tiny plane on the dial tilts to the right; if the pilot banks left, the tiny plane tilts left. Now, on *most* artificial horizon dials, if the pilot points the nose of the plane downward, the tiny plane drops below the horizon line on the dial. But exactly the opposite happens with the type of artificial horizon—a Sperry F3—that was installed in Jerry Dwyer's Bonanza. The Sperry F3 uses a weighted sphere, about the size of a billiard ball, to represent the earth, and when a plane equipped with a Sperry F3 dives, the ball stays steady and the plane pivots nose-down around it . . . causing the tiny plane on the dial to *appear* to be climbing higher in the sky. A pilot who was even slightly unfamiliar with the Sperry F3 (as the investigation concluded Peterson was) could easily misinterpret a dive as a climb—and could react by putting his plane into an even steeper dive. It's hard to explain how counterintuitive the instrument's climb/dive reading is, but Jon showed me a Sperry artificial horizon like the one in Jerry Dwyer's plane—he bought one on eBay—and when Jon simulated a dive with the instrument, I could have sworn I was in a climb. So it's reasonable to assume, from the accident summary, that Roger Peterson thought he was in a steep *climbing* turn when he was actually in a steep *descending* turn.

But not everyone believes the crash report summary. One of those who doesn't is Bill Griggs, the music historian. Griggs has done a lot of research on the plane crash and the government's investigation; he's read hundreds of pages of government documents resulting from the crash investigation, and he's talked with Jerry Dwyer, whose plane Roger Peterson flew into the ground that night. According to Griggs, Dwyer insists that Roger Peterson was quite familiar with the Sperry F3 artificial horizon. Dwyer also insists that the sky was clear—stars visible overhead, not clouds at 3,000 feet, as the report says. And the Big Bopper's son says Dwyer once told him that Peterson must have been "incapacitated in some way" to have crashed.

Incapacitated in *what* way? Again, here's where things get interesting. One possible mode of incapacitation came to light two months after the crash. As farmer Albert Juhl was doing his spring planting, he came across a loaded pistol in his cornfield at the site of the crash. According to a newspaper account at the time, Juhl was curious to see if the gun would still fire, so he pointed it skyward and pulled the trigger several times. Luckily for Juhl, the bullets exited the barrel cleanly, rather than getting stuck in some rust and causing the gun to explode in his hand. But what was the gun doing at the crash site? And was Juhl the only one to fire it?

The pistol, it soon emerged, belonged to Buddy Holly. He carried it, concealed beneath a false bottom in his overnight bag, for protection, supposedly, since he sometimes got paid in cash for his band's gigs. So now the story includes not just a plane crash and three dead celebrities, but a loaded pistol thrown from the wreckage. Not surprisingly, after the gun's discovery, a rumor began making the rounds that the pilot's seat had a bullet hole in it. Now, it doesn't make much sense that Buddy Holly, Ritchie Valens, or the Big Bopper would want to shoot the guy flying the plane, but rumors, no matter how outlandish, often spread like wildfire. The thing is, the pilot's body was autopsied—Peterson was the only one of the four autopsied—and there was no gunshot wound. Well, then, if the pilot wasn't shot, maybe somebody else was; that's how the rumor mill works.

Buddy Holly's wife Maria was pregnant at the time of the crash; soon afterward, she had a miscarriage. J. P. Richardson's wife was also pregnant; two months after the crash she gave birth to a son, whom she named Jay Perry. Jay Richardson grew into the spitting image of his father, in size, appearance, and even mannerisms. He also grew increasingly curious about the Big Bopper's life, music, and death. Jay's mother never talked much about his dad, so Jay began talking to other people who could fill in the blanks for him. And the more he learned, the more eager he became for answers to two haunting questions: Unlikely as it seemed, was it possible that the Big Bopper had been shot aboard the plane? And could he have survived the crash, however briefly, and died while trying to seek help?

I went to Beaumont hoping to give Jay Richardson the answers to those two questions. As it turned out, I think he got considerably more than either of us bargained for.

March 6, 2007, dawned clear and unseasonably warm in Beaumont. I'd landed in Houston about 10 P.M. the night before, and checked into the Hampton Inn in Beaumont a little before midnight. At 7 A.M. the following morning, I was at Forest Lawn Cemetery and Memorial Garden, where the Big Bopper's grave was marked by a silver historical marker, a horizontal bronze plaque, a green tent, some blue privacy tarps, and a yellow backhoe. Jay Richardson was already there, his silver Chevy pickup parked a short distance from the grave. At the entrance to the cemetery, a local television news crew had set up a camera and tripod, but their view of the grave was blocked by the tarps, which were fastened to the tent poles to screen the proceedings from curiosity seekers.

The grave directly alongside the Bopper's had been excavated the previous day. That one held Jay's mother. He was moving both parents to a new part of the cemetery so he could mark their graves with more prominent markers than the simple horizontal plaques required in this part of the memorial garden. A Texas sculptor is working on a life-size bronze statue of the Big Bopper, which Jay wants to put alongside his dad's grave. When Jay first began to make arrangements to move the graves, he realized that he'd never have a better chance to seek answers to the two lingering questions he had about his father's death.

The backhoe started digging by 7:30, and by 7:35—just three or four feet down—I was starting to worry. During one of our phone calls, I'd asked Jay some questions about the cemetery. It was in a low-lying area, he'd said, with lots of water just under the ground. When I heard that I thought, "Well, that doesn't bode well for the preservation of skeletal remains." Sure enough, as the dirt came out, scoop by scoop, water rushed in to take its place. The cemetery crew had foreseen this problem, and had a gasoline-powered pump standing by; they fired it up, and it quickly removed most of the water from the open grave. But as soon as they shut the pump off, water began refilling the hole.

Most of Beaumont, remember, is only a few feet above sea level, so the water table isn't far belowground. If the casket had filled with water, it was possible the

Bopper's remains would be down to bare bones—or even decayed to mere fragments of bone. If the soil here were as acidic as Tennessee's red clay, forty-eight years of marinating in groundwater could easily dissolve the soft tissue and much of the bone.

As I watched the excavation of the Big Bopper's grave—a watery grave if ever there was one—all I could do was hope that the burial vault surrounding the casket had worked the way it was supposed to. Burial vaults come in two basic designs, reflecting two basic materials. One type is a rectangular box, several feet deep, made of concrete and capped with a flat concrete lid, sealed with a rubber gasket. The National Concrete Burial Vault Association (NCBVA)—yes, there really is such an organization—has one article on its web site titled "Why a Concrete Burial Vault?" (because it's really strong, is what the answer amounts to) and another titled "Here's How Concrete Vault Protection Withstands Different Types of Loads" (by being really strong, is basically how). What the NCBVA's web site doesn't say is that concrete vaults can leak. If the water table rises up to the seam where the lid meets the box, and if the seal's not perfect, the vault can be transformed from a water-excluding barrier into a water-filled vessel.

The other type of burial vault is steel, and it's sealed not at the top but at the bottom. Sealing it at the bottom seems odd, but there's method to the apparent madness. To simulate the way a steel vault keeps water out, take a clear drinking glass, turn it upside down, and submerge it completely in your kitchen sink. Water will rise inside the glass as you begin to submerge it, but then—in my own kitchen-sink experiment, after rising only a fraction of an inch—it stops, held at bay by the air pressure inside the glass. A diving bell works on this same principle. So the advantage of a steel vault is that it doesn't rely on a perfect seal or gasket to keep water out; it relies on a simple principle of physics: as long as the air pressure is greater than the water pressure, the air keeps the water out. (A page on the web site of Clark Burial Vault, which makes steel and copper vaults of this design, is titled "Metal Versus Concrete"; a cutaway drawing depicts a dry, air-filled metal vault and, beside it, a concrete vault completely filled with water.)

But nothing's perfect, and steel burial vaults are no exception. They have two Achilles' heels: if a cemetery floods, steel vaults—being lightweight and air-filled—can bob to the surface and float away, bearing the dearly departed downriver for miles. That's what happened with several hundred vaults (steel ones and even some concrete ones) after Hurricane Floyd flooded Tarboro, North Carolina, in 1999, and with more than 1,200 in

Louisiana when Hurricane Katrina hit. The other occasional problem with steel vaults is that corrosion or damage can create a hole in the dome-shaped top of the vault, and if that happens, all bets are off: water is free to rise within the vault because the hole allows air to escape rather than becoming pressurized.

The Big Bopper's casket had been sealed in a rust-resistant steel vault made of galvanized steel one-seventh of an inch thick. As the backhoe scoop dug down past the three- and four-foot marks, down toward the lower edge of the vault, it was clear that the vault had been at least partially submerged for much of the past forty-eight years, and it seemed likely that at times it had been completely submerged. I hoped there was no pinhole anywhere in the dome that would have allowed water to rise within it and submerge the coffin. But we couldn't make even an educated guess until we removed the top of the vault, and we wouldn't know for sure until we opened the casket and saw whatever was left of the Big Bopper.

Once the vault was fully exposed in the soggy hole, the workers—led by a short, lively foreman nicknamed "Fish," who once toured as a rock-and-roll musician (inspired, in his youth, by the Big Bopper)—slung steel cables around both ends, and the backhoe operator hoisted the vault out of the soggy hole and onto a small trailer. We then got into our vehicles and drove from the grave with the body—a funeral procession in reverse—to the cemetery's maintenance shop, a large metal building tucked into a stand of pine trees at the far side of the grounds. Cleaning the mud off the vault took nearly half an hour, and figuring out how to open the vault took another ten minutes or so, as there were no obvious bolts or clamps at the corners or on the sides. Finally someone spotted two small levers, practically invisible, recessed into each end of the vault. Using a mallet and a large screwdriver, Fish carefully tapped the levers sideways, and the dome separated slightly from the base of the vault. Again the cables were snugged onto the vault, and the backhoe hoisted the dome straight up and swung it off to the side, revealing a silvery-gray casket with a low pedestal base.

A dull film and a few traces of surface rust marred the casket's once-gleaming finish, and a pale, crusty line of mineral deposits partway up the pedestal showed that a couple of inches of water had seeped into the vault before the water pressure stopped its rise. But the waterline was well below the bed of the casket, and for the first time that morning—the first time since seeing the waterlogged grave—it seemed possible that the remains of the Big Bopper might be at least partially intact. A hearse was waiting to take the casket back to the funeral home, where I would examine the body or the bones in the embalming room.

But then the funeral home's manager, Rodney Landry, began to have misgivings. Even with the casket still sealed, the unmistakable odor of decomposition began suffusing the surrounding area. Rodney had the casket brought inside the maintenance building, where he removed a small rubber plug from a vent near the foot of the casket. The odor, already strong, became intense. Rodney shook his head. Gravely. "I can't bring this into the building," he said. "This smell would permeate the building for weeks." He was right, I knew from experience.

"I can just do the examination here," I said. "That's fine with me." Rodney conferred with Jay, who also understood the problem and was willing to be considerate of others. We would do the examination in the shop. By now it was nearly noon and I was hungry, so we made a quick run to Mr. Burger, a tiny carryout place a mile down the road, and wolfed down lunch standing outside the cemetery shop. We positioned ourselves upwind from the coffin, but occasionally the wind would shift and the aroma of burgers would mingle with the aroma of the Bopper.

As soon as we finished eating, I was ready to open the casket and get to work, but Rodney stopped me.

"Before we go any farther, we need everyone here to sign a hold-harmless release," he said. Normally these are meant to shield a business from liability if a visitor or a customer—for example, a student pilot, or a participant in a whitewater rafting trip—gets hurt or killed. In this case, the form was meant to protect the funeral home from any sort of lawsuit alleging emotional trauma from the potentially gruesome sight we were about to see. I knew I wasn't going to be traumatized by whatever lay inside the coffin, but I could understand how the sight of a skull, its flesh half rotted away, might traumatize Jay Richardson or historian Bill Griggs or one of the few other people allowed to witness the proceedings. Everyone, including me, gathered in a circle to sign the required form, and then Randy Steele—a friend of Jay's—said a brief prayer, asking God's blessings on Jay and everyone else present. After Randy and a few of the other people circled up had murmured "amen," Rodney gave a nod and Joni Johnson, a funeral director and embalmer, inserted a small metal handle near the foot of the casket. She cranked it clockwise, releasing the latch at that end, then repeated the maneuver at the head of the casket. Both halves of the hinged lid were now free to swing up and open.

The vault and coffin had performed well; now, the crucial question was, How well had the embalming held up? I still wasn't optimistic. Occasionally someone donates an embalmed body to the Body Farm, and I've seen them decay—not as swiftly as unembalmed bodies, which can skeletonize in two weeks or less at the height of summer, but sometimes in a matter of months. Most funeral homes do what's called "cosmetic embalming," which is intended mainly to keep a body looking and smelling good for the brief period required to arrange and conduct a memorial service. The procedure is this: the embalmer inserts a catheter into the carotid artery, in the neck, and begins pumping embalming fluid—a mixture of formaldehyde, ethanol, and other solvents—into the circulatory system. Another catheter, inserted into the jugular vein, drains blood from the body to make room for the embalming fluid. To help distribute the fluid, the embalmer massages the body, kneading the liquid through the smaller blood vessels and out into the muscles and other tissues. If the funeral service involves an open casket and a viewing of the body, sometimes the embalmer injects small amounts of fluid under the skin of the face as well. If the funeral home is doing a thorough job, the embalmer will also embalm the body cavity, inserting a trocar (a hollow cylindrical spike) that allows the organs to be removed by suction aspiration and the cavity to be filled with embalming fluid. A plastic trocar "button" is then screwed into the trocar abdominal opening to seal it. That, as I say, is a thorough job; judging by some of the corpses I've exhumed or seen decay at the Body Farm, not every embalming job is anywhere close to thorough. One series of slides in my files shows an embalmed man lying on the ground at the research facility; the man's upper body remains relatively intact, but his legs are virtually skeletonized, an indication that very little embalming fluid was worked into his limbs.

More than a century ago, Abraham Lincoln received a very thorough embalming, and his body reportedly held up quite well during the nearly three weeks that elapsed between his assassination and his burial. What's more, Lincoln's body remained in surprisingly good—and easily recognizable—condition thirty-six years later, when it was exhumed so it could be placed beneath a larger, more secure monument. Civil rights activist Medgar Evers, shot in Mississippi in 1963, likewise had a great embalmer, his body remaining well preserved enough for an autopsy to be performed when it was exhumed twenty-eight years after his murder. (African-American funeral homes tend to do particularly good embalming, I've noticed.) Celebrity Anna Nicole Smith's body, on the other hand, decayed rapidly even after she was embalmed, probably because the ten-day delay in embalming gave the internal bacteria a head start on the decomposition process, despite the fact that her body was kept in a morgue cooler. All this is by way of background so you'll understand my uncertainty and concern about the sight that awaited us inside the Big Bopper's casket.

Jay Richardson stepped up beside the upper end. I stood beside him. A hush fell over the group as Rodney Landry grasped one edge of the lid and gave a tug. After nearly half a century in the soggy ground, the lid raised smoothly and silently. As it did, it revealed one of the most remarkable sights I've ever seen in my career. What I saw, and what Jay Richardson saw, was not a bare skull or rotting tissue. What we saw was the Big Bopper himself, complete with his trademark crew cut. The skin was discolored—a mottled bluish-purple—but the features were instantly recognizable, and bore a striking resemblance to those of the man who stood beside me, getting his first glimpse of his father in the flesh.

The body's condition was all the more remarkable in view of the damage it must have suffered in the plane crash. Embalming fluid is volatile, and it gradually leaches out of the body, even if the skin has no breaks. In cases of extensive trauma, trying to fill a body with embalming fluid is like trying to fill a sieve or a sponge. To help lessen the leakage, the funeral home had zipped the body inside a "Unionall," a leak-resistant jumpsuit, before dressing the Bopper in his funeral suit.

For Jay's sake, I was relieved at the body's excellent condition. But for me, it complicated things. One of the questions haunting Jay—could his father have survived the crash long enough to walk away and climb the fence that ran between his body and the plane's wreckage?—would have been easiest to answer by examining bare bones, whose fractures (or lack of fractures) would have whispered the story of the Bopper's death to me, bone by bone. I could have seen and felt the details directly, then shown and explained them to Jay one by one. Every fracture tells a story, and if you study it closely, you can often piece together a surprisingly detailed narrative. A blow to the skull, for instance, tends to produce fractures that radiate outward from the point of impact. If the skull receives several blows, it's sometimes possible to tell the order in which they occurred, because if the fracture lines from two blows intersect, the fracture from the second blow will usually stop when it reaches the first crack, rather than jumping across that gap and continuing.

I hadn't counted on a nearly intact corpse; to remove all of this tissue in order to examine the bones would be difficult and time-consuming. It would likely take several days to remove the tissue, simmer the bones, and scrub them clean. It would also prolong Jay's emotional turmoil, and probably increase it dramatically, too, to see the recognizable body of his father transformed into a bare skeleton. Besides, the funeral home wasn't equipped to deflesh bodies, only to preserve them. Back in Knoxville, I'd have had no difficulty cleaning off the Bopper's bones; the Regional Forensic Center, housed in the morgue at UT Medical Center, has a "decomp room" designed specifically to speed the process, with oversized sinks, steam-jacketed kettles for simmering bones, and heavy-duty exhaust fans. But the situation was entirely different at Forest Lawn, and if they didn't even want to open the casket in the building, they certainly wouldn't want me to spend two or three days marinating the structure in the odor of decomposition.

Fortunately, we had a Plan B: a portable X-ray system provided by MobilexUSA, a nationwide radiology service. Cindy Solomon, the manager of Mobilex's Southwest regional office, had made the 300-mile trek from Dallas to help shoot X-rays of the body, assisted by Joey Marse, one of the company's radiologic technologists. Besides the mobile camera, they'd brought a couple dozen cassettes loaded with film so we could shoot multiple images of the entire body if we needed to. They'd sealed the cassettes in plastic, which would protect them against contamination without interfering with the images. These Mobilex folks were clearly no slouches.

While we'd been getting the vault cleaned up and the coffin opened, Cindy and Joey had been setting up the X-ray machine in the funeral home's embalming room. Now that we'd decided not to do the examination there, they scurried across the street to pack up their gear and bring it to the shop. As soon as they returned and unloaded it, though, another wrinkle occurred. "Where's the closest 110-volt outlet?" asked Joey.

Rodney turned toward a workbench along one wall, then his face took on a stricken look. "A truck hit the corner of the building on Friday," he said, "and ripped the power line loose."

"So there's no electrical power here?"

Rodney shook his head sadly. Someone asked if the shop had a generator, and Rodney brightened.

Now it was Joey's turn to look downcast. "The system can't run off a generator," he said. "The current fluctuates too much."

Rodney huddled with his staff for a moment, then asked if we could take the X-rays in the funeral home's garage, which had electrical outlets and—perhaps equally important from Rodney's perspective—plenty of fresh air and well-sealed doors separating it from the building's interior. I said that was fine with me, and

Jay agreed, so Cindy and Joey repacked their gear yet again and we loaded the Big Bopper's body—wrapped in a plastic tarp, to contain any fluids—into a gleaming black hearse for the 200-yard procession to the Forest Lawn garage.

There, finally, we got to my reason for being in Beaumont. It was only early afternoon, but I'd been at the cemetery and mostly on my feet for eight hours already, and I was getting tired. But now that I was setting to work, I felt my adrenaline kick in. What was I about to see in the skeletal X-rays, and would it be sufficient to answer Jay's questions?

With the body laid out on a waist-high table, Cindy, Joey, and I quickly got into a rhythm. We started at the top, sliding a cassette beneath the pillow that supported the head. After making sure everyone was at least six feet away, Joey snapped the exposure. Then Cindy removed the cassette, handed it to Joey, and helped me position the next one, this one beneath the chest.

The Bopper wasn't called "Big" for nothing: the smaller cassettes weren't big enough to image his whole head, and the larger ones didn't quite span the width of his rib cage and pelvis. I was glad Cindy had brought plenty of cassettes.

We shot a series of images covering the entire body, which Joey drove across town to MobilexUSA's

Beaumont lab. While we waited for him to process the films and return, we dispersed in clumps across the parking lot behind the funeral home, some of us sipping soft drinks in our vehicles, some sitting or lying in the soft grass in the warm Texas sun. Rodney and Joni sat in wheelchairs inside the garage, the lights switched off. Garbed in disposable white gowns and masks, they looked vaguely sinister, like mad scientists or villainous doctors from some sci-fi horror film.

Joey was gone for about forty-five minutes. His return was a call to reconvene, and people struggled up from the grass and out of their cars to see what stories the X-rays would tell. Answering one of Jay's questions—had his father been shot?—required a look at all of the films, one by one, on the portable light box Joey set up in one corner of the garage. There are two different ways to tell, from X-rays, if a person has been shot. If the bullet hits bone, it will either make a nick or punch a hole in the bone. If it enters the skull, it leaves a hole; if it hits a long bone, like an arm or a leg, it could leave either a nick or a hole. Either way, as the bullet enters or crosses the bone, lead rubs off the bullet onto the bone, leaving what's called a lead wipe around the hole or along the nick in the bone. These smears of lead—which X-rays can't pass through—show up as streaks of bright white on the films. Another sign that someone has been shot is the presence of little white specks in the X-ray images. When bullet hits bone, fragments of lead peel off the bullet and either splatter against bone or lodge in the soft tissue. The X-rays of the Big Bopper clearly showed pieces of metal, but they were clearly zippers—the zippers of the Unionall his body wore to minimize fluid leakage from his many crash-related wounds. Nowhere, amid the bones or the soft tissue, did I see the bright white patches characteristic of lead smears or fragments.

I looked at Jay. "There's no evidence of any foul play," I told him. "No sign of a bullet anywhere in the body. I think we can lay that rumor to rest now." He nodded.

Jay's second question was whether his father might have survived the crash and tried to go for help. To answer that, I led Jay, film by film, through the injuries his father had sustained in the crash. They ranged, literally, from head to foot. In his right foot, all five tarsal bones (the bones fanning out below the ankle) were broken, in a line straight across the mid-foot—a type of fracture often called a "march fracture," because it's common in soldiers, who do a lot of marching on hard-surfaced parade grounds and on field exercises, carrying heavy packs.

Both lower legs had compound fractures of the tibia, the larger of the two bones in each lower leg. The left tibia actually had three fractures: one at about mid-shaft, another in the distal (lower) third of the bone, and a third one at the line where the bone's shaft joined the distal epiphysis, the end of the bone. In the right lower leg, both the tibia and the fibula had compound fractures just below the knee; in addition, the lower end of the right tibia may have been fractured at the epiphyseal line, as the left tibia had been. I'd already suspected there would be severe fractures to the legs, because of the odd angles at which the feet pointed out toward either side of the casket.

The upper legs were also shattered. The left femur had a compound fracture slightly above mid-shaft; the right femur was destroyed at the knee: the distal end of the femur was broken away from the shaft, and the condyles—the two knuckle-like knobs at the end of the bone—were broken apart.

The pelvis, too, had been shattered. The left innominate (hip bone) was broken apart at the sacroiliac joint, where the wide, flaring part of the bone joins the sacrum, the bottom of the spine. In addition, the left pubic bone was snapped off, as well as torn free of the pubic symphysis, where it had once met the right pubic bone at the midline of the pelvis. The break in the left hip was so severe that the entire left leg was rotated sharply to the outside.

"I think I can say with great certainty that there's no way your father walked away from that crash," I told Jay. As he studied the images of devastation, I could see that he understood and believed me.

The chest was crushed; multiple ribs on the right and left side were broken, and the left clavicle (collarbone) had two compound fractures. The spinal column appeared to be broken in two places. One definite fracture was near the bottom of the chest, between ribs nine and ten. It also looked as if the spine might have fractured in the neck, just below the skull, between the first and second cervical vertebrae. I suspected this might be a deceleration fracture, the type that results when the body is stopped abruptly (by whatever crushed the Bopper's chest) while the head continues forward. Cindy Solomon called it a "hangman's fracture" (a term I hadn't heard before): the same abrupt deceleration occurs when a noose jerks the head to a halt, while the momentum of the body continues downward. I reckon the reason Cindy knew that term and I didn't is that Texas is a hanging state but Tennessee isn't: for crimes committed after 1998, lethal injection is the prescribed method of execution, while people who are on death row for older crimes get to choose: the needle or the electric chair. (Later, I asked Dr. Ken Rule, a UT radiologist, to review the Bopper's X-rays with me; Ken didn't think the neck had been fractured.)

The Bopper's skull had numerous fractures. When I studied the X-ray of the head, I was even more impressed with the work the funeral home had done forty-eight years before. The bones of the face were largely crushed—I'm not sure what was supporting the face of the corpse, but it wasn't bones. The right side of the forehead was crushed, as was the right parietal bone, just above the forehead. A large fracture had split the juncture of the right parietal and right temporal bones, and another large fracture bisected the skull at the midline, along the sagittal suture (at the junction of the two parietal bones, which form the sides of the skull).

It would have been impossible for his father to walk away from the plane's wreckage and climb a fence, I told Jay; he had to have been thrown clear. He had so many fractures—fractures to the skull, fractures to the chest, fractures to the pelvis, fractures to both legs—he could not have walked. In fact, I said, he probably died almost instantly from the massive fractures to the skull and chest. Conceivably he could have survived the shattering impacts to his legs, but not such trauma to the head and chest. Again Jay nodded thoughtfully.

Although Jay didn't uncover a conspiracy surrounding his father's death, I think something else occurred when that coffin was opened: it gave Jay Richardson the chance to visit his dad. The Big Bopper, remember, died about two months before Jay was born, so he'd never seen anything but photographs. As we worked all day in Beaumont, I watched Jay, and I think he felt much better than before. Psychologically, I think it was a great day for him, a turning point.

A person who has lost a parent early in life—or, as in Jay's case, even before being born—doesn't know what that parent was like. I lost my own father very young; he committed suicide when I was only three, and as I've gotten older, I've found myself thinking more and more about him, wondering about the relationship we might have had. During the course of the exhumation and examination, I told Jay about losing my father. I wish I'd had the opportunity to meet my dad the way Jay had—it might have shed at least a little light or filled in some of the blanks in my knowledge of my father. Some of my lingering questions are unanswerable, of course, such as why my father felt such despair that he shot himself, but at the very least, I'd have liked the chance to say goodbye to him, face to face. I think Jay bonded with his father in a new way, and I think he felt better at the end of the day than he'd felt for all those years he'd been wondering.

Back when we first began planning this, I worried that if I just wrote a report, Jay might still have questions—he wouldn't have physically seen what I was talking about, and I might not have given the best possible explanation. That's why I'd asked him, well ahead of time, if he thought he might be able to be present during the autopsy, and might be able to assist me. By the time I arrived in Beaumont, Jay had prepared himself emotionally to observe the entire examination, and at the end of the day, we were both glad he had. There would be no lingering doubts, at least as far as what I could show him and explain scientifically.

I've consulted with families for many years on the death of loved ones; I'm often the last person in the chain of identifying people. And often it's extremely painful for them, especially if the body's in bad condition. But the Big Bopper's body had been in remarkable condition, allowing Jay to recognize his dad, to gain a better sense of where he himself had come from. Jay later told the *Beaumont Enterprise* the experience was "one of the grandest days of my life."

Not only did Jay see his father; Jay's two sons and his daughter got to see their grandfather. The two sons had a hard time at first, but when all was said and done and we were saying our goodbyes, Jay's older son, who's a teenager, came up and hugged me. At that moment, I realized that I'd given him the only opportunity he would ever have to see his grandfather.

This wasn't a criminal case, but it was a mystery: a haunting personal mystery for the family. It was also a lingering mystery for many fans of the Big Bopper. We were able to solve that mystery, lay those questions and rumors to rest. And lay the Big Bopper himself finally to rest.

Jay Richardson and I went to Beaumont to dig up answers about a performer and a father who died almost half a century ago. And we found them, thanks to the chemistry of embalming, the physics of air pressure, and the illuminating insights of X-rays.

16

LEOMA PATTERSON, PART 3: PUSHING THE LIMITS OF DNA TESTING

The airport security screener peered closely at the two small objects in the envelope; then he called over another screener, and another, and a supervisor. Within minutes I was surrounded by armed Transportation Safety Administration guards, questioning me closely about the two human teeth I'd been carrying in my pocket. Luckily, I was on my home turf—Knoxville's McGhee-Tyson Airport—and half the guards were familiar with me. They weren't suspicious, merely curious, and after I'd explained why I was flying to Texas with two teeth from a mysterious woman's skull in my pocket, they waved me on and wished me luck finding an answer to the question that had bedeviled me for months in the Patterson case: Was this Leoma, or was this not Leoma? I wouldn't find the answer itself in San Antonio, but I would meet the scientist who—if skill and luck converged—could give me a definitive answer. Like a runner carrying the flame from Athens to the Beijing Olympics, I was passing the torch to the next person in this scientific run for the truth. In this case, the torch consisted of teeth, and the "flame"—at least, what I hoped I was handing off—was mitochondrial DNA.

Millions of people first learned of the forensic potential of DNA during the highly televised murder trial of O.J. Simpson. Blood drops leading from the spot where Nicole Simpson and Ronald Goldman were slaughtered matched O.J. Simpson's blood type and DNA, a parade of prosecution witnesses testified. What's more, blood matching Nicole's DNA was found on a pair of socks in O.J.'s bedroom, and blood matching Ron Goldman's DNA was found in O.J.'s Bronco. The odds against anyone other than O.J., Nicole, and Goldman being the sources of those bits of bloodborne DNA were astronomical, ranging as high as nearly one in ten billion.

The reason for such overwhelming mathematical probability, the jury learned during days of mind-numbing scientific testimony, is DNA's immense size, coupled with its consistent structure and deceptively simple building blocks. Its now-famous shape, called a "double helix," resembles a ladder whose uprights have been twisted so they spiral or corkscrew around one another. The three billion "rungs" of this corkscrew ladder consist of pairs of chemical bases named adenine, guanine, cytosine, and thymine (abbreviated A, G, C, and T). For biochemical reasons I won't go into (because I don't begin to understand them!), every "rung" with an A on one side always—no exceptions—has a G on its other side; similarly, C is always paired with T. So if you were magically reduced to the size of an atom and you began climbing this three-billion-rung ladder, the first five rungs you ascended might be labeled this way: AG, AG, CT, GA, TC.

Statisticians have an old saying about monkeys and typewriters that's meant to illuminate how randomness, given enough time, produces specific meaning; one variation of that saying goes like this: Chain a hundred million monkeys to a hundred million computers for a hundred million years, and one of them will write the complete works of William Shakespeare. Well, I don't actually believe that—I suspect it's far more likely that they'll bring the Internet and the entire computer-dependent world to its knees, and in far less than a hundred million years. But I can believe this, and can grasp it at a very rudimentary level: Stack up three billion pairs of AGs, GAs, CTs, and TCs, one atop another—or string three billion beads on a genetic necklace, or strike three billion two-note chords on a piano keyboard—and you're going to get one-of-a-kind variations. Variations named O.J. Simpson. Nicole Simpson. Ronald Goldman. Bill Bass. And Leoma Patterson. All it takes to match a forensic DNA sample to a particular individual with immense certainty is a good forensic sample, a good sample from the individual, and impeccable collection and handling procedures. In the O.J. trial, the defense didn't challenge the scientific validity of DNA analysis, the matches, or the statistical probabilities. Instead the defense claimed—and the jury seemed to believe—that the incriminating samples were either planted by detectives or contaminated by careless lab technicians. So despite the staggering odds of guilt the prosecution and its experts invoked—billions to one—O.J. walked. The case hadn't been made beyond a reasonable doubt.

What about Leoma? Could a new DNA test indicate, beyond a reasonable doubt, whether the remains I'd exhumed, not once but twice, were hers or some other murder victim's? No one was on trial for murder here—Leoma's great-nephew confessed in 1985 to killing her and served six years for manslaughter, though in 2006 he retracted his confession. It wasn't a jury but a family whose questions I hoped to answer beyond a reasonable doubt.

Unfortunately, my chances of getting a good DNA sample—a fresh DNA sample—were a lot slimmer than the Los Angeles Police Department's chances at a fresh crime scene. The DNA molecule that was the centerpiece of the LAPD's analysis and the prosecution's expert testimony was nuclear DNA, which takes its name from its address: the nucleus of every cell in the body. Besides being immense, nuclear DNA is also relatively fragile, at least over the years that often intervene between the time someone is murdered and the time I examine their bones. Gradually, bacteria, acids, and other chemicals in the body and the environment consume or corrupt the long, looping strands of nuclear DNA. Probably my best hope was that a bit of mitochondrial DNA—mDNA, a shorter, simpler, but hardier cousin of nuclear DNA—had survived inside the tough, enameled capsules of the two molars I'd carried at 34,000 feet from Knoxville to San Antonio.

San Antonio's Riverwalk, packed cheek by jowl with restaurants, bars, hotels, and tourists, is like a miniaturized, Disneyfied version of the canals of Venice or Amsterdam. Sprawling alongside the Riverwalk, the Henry B. Gonzales Convention Center encompasses 1.3 million square feet, or about 30 acres—about the size of a small farm in the hollows of East Tennessee. Toward the end of February 2007, those 1.3 million square feet were swarming with thousands of forensic scientists converging on San Antonio for the annual convention of the American Academy of Forensic Sciences (AAFS). The AAFS boasts more than 6,000 members, in fields ranging from A (anthropology) to Z (zoarchaeology: "Doc, are those bones animal or human?"). The convention center's cavernous exhibit hall—bigger than the football field in Neyland Stadium—was jammed with hundreds of vendors' booths hawking everything from DNA analysis, fingerprint scanners, and toxicology screenings to disposable biohazard suits and rape-test kits.

At booth number 117—sandwiched between two software companies—I met Jason Eshleman, a slight, soft-spoken, but swift-talking scientist. Jason is one of an emerging crop of twenty-first-century anthropologists, inhabiting a niche I couldn't have imagined a half century ago: molecular anthropology.

Jason earned a Ph.D. at the University of California at Davis, under the tutelage of David Glenn Smith, who helped pioneer the application of modern DNA testing to ancient human skeletons. Smith enticed Jason to UC-Davis for his graduate work by posing an intriguing anthropological question he was creating a research project to answer: Were the dominant Native American tribes of the American Southwest—the Apaches and Comanches—the original human inhabitants, with a claim on the land dating back some 10,000 years or so, or were they (as Smith believed) interlopers, who'd invaded from Canada relatively recently—"only" about 4,000 years ago? Smith told Jason that DNA could give a definitive answer; what he didn't tell Jason was that extracting and purifying ancient DNA was a nightmarishly difficult process. "I wouldn't wish it on my worst enemy," Jason now laughs, post-Ph.D.

Trace Genetics, the DNA lab Jason had cofounded after earning his Ph.D., had recently been acquired by DNA Print, a much larger firm, because of Jason's expertise in detecting and analyzing ancient DNA. One of Jason's most notable successes had been finding DNA—and plenty of it—in the teeth of a 10,000-year-old skeleton from the Queen Charlotte Islands, located off the coast of British Columbia (an archipelago that bills itself as "the Galápagos of Canada"). Encouraged by this, I handed over the envelope containing two molars I'd pulled from the jaws of Maybe-Leoma.

Some scientists thrive on the conceptual; their minds can envision particles that the most powerful microscopes can't show us; processes that can't be directly observed, but only inferred, guessed at, by interpreting a stew of complex biochemical by-products. I am not one of these scientists; I need bones and teeth, things I can see with my eyes and grasp with my hands. Jason Eshleman, on the other hand, can see with his mind's eye, grasping the complex interactions of the most complex molecules in the body, DNA.

One of the most famous, most challenging, and most controversial samples Jason had worked with was Kennewick Man, a male skeleton a college student stumbled upon in the shallows of the Columbia River near Kennewick, Washington, in 1996. Initially believed to be the remains of a white homesteader, the bones were sent to a carbon-dating lab just to be sure. The results—which put the age of the bones at 9,200 to 9,600 years old—ignited a fierce controversy. On one side were Native American tribes who wanted to rebury the remains as quickly and reverently as possible; on the other were scientists who wanted to study the bones for more clues about who first settled the Americas, where they came from, and how they got here. Several of my former Ph.D. students got caught up in the case, including George Gill, who teaches at the University of Wyoming; Doug Owsley, who heads the physical anthropology division at the Smithsonian Institution (and who got his bachelor's degree at Wyoming, under George Gill); and Richard Jantz, who directs the Forensic Center at UT (and who developed ForDisc).

Just months after the discovery of Kennewick Man, while Jason was still working in David Glenn Smith's lab, they received a small sample from the ancient bones for DNA testing. Before they had a chance to process the sample, though, the FBI confiscated it. A consortium of Native American tribes had filed a lawsuit in federal court, and the judge swiftly granted an injunction to block the DNA testing and retrieve the sample. Smith's lab handed over the sample, but not before placing it in a vial with a tamper-evident seal.

Nine years later, after scientists won the right to make a thorough study of Kennewick Man, Jason got the sample back—still sealed in the vial. Unfortunately, the only DNA in the sample proved to be quite a bit younger than 9,000 years; it came from another graduate student who'd worked in the lab back in 1996. Jason wasn't able to learn the genetic makeup of Kennewick Man, but he did learn firsthand how easily ancient samples can be contaminated. It was a lesson underscored on multiple occasions by a mysterious, persistent contaminator. "There was a period when I kept seeing the same DNA sequence over and over in my samples," he recalls. "It wasn't from me, and it wasn't from anyone else who worked in the lab—we had everyone's sequence on file, so we knew it wasn't any of us." Eventually the culprit emerged: Jason's girlfriend, who shed enough stray skin cells to make her presence known, even though Jason faithfully scrubbed up and suited up, much like a surgeon, before entering the lab.

As we talked amid the thicket of vendor booths in the cavernous exhibition hall in San Antonio, Jason explained how he would seek out whatever mDNA the teeth contained. He would start with just one of the teeth, keeping the other in reserve. His first step—and the reason he hadn't worried about my contaminating the teeth with my own DNA—would be to scrub the tooth with a solution of bleach to remove any dirt or other residue, then soak it in the bleach solution another five minutes. The bleach would destroy any DNA on the outside of the tooth, but five minutes wouldn't be enough time to allow the bleach to penetrate the tooth's enamel and permeate the dentin, where the mDNA was ensconced. Next, he'd rinse it with sterile water, then dry it under an ultraviolet lamp (another step designed to destroy any contamination on the surface). To gain access to the DNA, he would then crack the tooth into smaller fragments and soak them in a solution that would dissolve the tooth's calcium, break down its proteins, and free the DNA from the cells. Once the pieces were dissolved—after about a week, he said—he'd bind the DNA to silica beads, extract the digested proteins and other grunge, and finally wash the beads with an alcohol solution. "Just as alcohol washes away people's inhibitions," he explained, "it relaxes DNA's inhibitions, causing it to release from the silica beads."

That's when the real biochemical chemical wizardry would begin. Heating the solution almost to boiling causes the two legs of the DNA ladder to peel apart—like a zipper unzipping. When that happens, short (twenty-rung) pieces of complementary DNA called "primers" latch onto the long, unzipped halves of the mDNA ladder wherever their own sequences of As, Gs, Cs, and Ts mesh with the mDNA's—a step called "annealing." The primers keep the ladder from zipping back together again as the solution is cooled slightly. Then, with the DNA unzipped, an enzyme called Taq (rhymes with "gack")—extracted from organisms that live in sulfur hot springs and hydrothermal ocean vents—moves in and builds a complementary new ladder-leg on each of the long unzipped legs . . . transforming one unzipped ladder into two ladders, creating two mDNA strands out of one. That entire cycle, called a polymerase chain reaction, or PCR, takes anywhere from twenty seconds to two minutes. At that point, the entire solution is reheated to near boiling again, the *two* ladders are unzipped, and the replication is repeated. It doesn't take many PCR cycles of unzipping and replicating—doubling the number of copies each time—to turn a few strands of mDNA into many; in fact, in an hour or less, a single strand can be transformed into billions of copies.

My head spinning from biochemical and genetic overload, I wished Jason good luck in his search for DNA in the teeth.

I did one other thing in San Antonio, while I was surrounded by forensic expertise: I showed a photo of Leoma Patterson, a photo of Joanna Hughes' clay facial reconstruction, and a freeze-frame from the video superimposition (dissolved halfway between the skull and Leoma's photo) to 106 of the scientists attending the conference. The question I asked them was simple: Is the woman in the photo the same woman shown in the clay reconstruction and the video superimposition, or is it a different woman? Of the 106 people I polled, 72 said, Yes, the same; 34 said, No, not the same. It wasn't a scientifically rigorous experiment, but it sure was interesting to see people struggle to decide, and to listen as they debated aloud with themselves or colleagues before voting. Later I bumped into Murray Marks—one of my former Ph.D. students who is now on the forensic faculty at UT (but whom I rarely see in Knoxville). Murray has used video superimposition, and has done research on computerized facial reconstruction, so I was particularly interested in Murray's opinion. He studied the video—not just the freeze-frame, but the actual video image dissolving from the photo to the skull, then back to the photo again—then glanced up from the computer screen. "What, are you retarded? Of course it's the same woman." I made a mental note to ask Murray to come along when I showed the video to Leoma Patterson's descendants. If, that is, Jason found DNA evidence in the teeth confirming that the bones were Leoma's after all.

I didn't realize what a big if that would turn out to be.

Three weeks after I handed the teeth to Jason, he phoned with some discouraging news. "I'm not getting any DNA from the sample," he said. The reason wasn't clear. "It might be that I'm getting a lot of interference from humic material," he said—various chemicals from the dirt, basically, junking up the reaction—"or it might be that there just isn't any DNA left." It was possible, he conceded, that acids from the soil, mold from the damp coffin, or bacteria had gradually invaded the tooth and destroyed the genetic material. I was dumbfounded; how was it possible for DNA to survive for 10,000 years in teeth in the Pacific Northwest, but not for 30 years in East Tennessee? "I don't know," said Jason. "There's not a lot of data from that part of the country." There wasn't much consolation in the idea that we were providing new, discouraging data. "There's an extraction I can perform to remove the humic acids," Jason offered. "It might also remove some of the DNA—if there is DNA—but it might leave enough behind for me to get a sequence." At this point, we seemed to have nothing to lose by trying.

A nail-biting week passed. "I'm seeing some DNA," Jason finally reported. Hallelujah! My excitement was short-lived, though: he'd identified some pieces of mDNA, but not enough yet to stitch together an entire sequence. "I'm going to process the second tooth now," he said, "and I'm hoping that will give me enough for a whole sequence." I hoped so, too.

But my hopes were in vain. The second DNA extraction failed to yield a complete mDNA sequence. So did a third, a fourth, and a fifth. I had sent Jason more teeth; I'd sent him a four-inch section of femur; I'd sent him a cheek swab from one of Leoma's granddaughters, Michelle Adkins, for comparison. But there was nothing in the skeletal material to compare. Back in 1979, it seems, Dr. Blake—the M.E. who did the original autopsy—simmered the bones in water containing lots of detergent and bleach. The chemicals had probably destroyed the DNA.

Finally, in late May—three months and seven fruitless extractions after I'd optimistically handed him those first two teeth in San Antonio—Jason said it was time to abandon our DNA quest. That meant abandoning our hope of answering the question still haunting the Patterson family: Was it Leoma, or Not-Leoma, who'd been buried in that hillside grave? Ironically, after nearly two years, countless hours, and the best forensic techniques we could apply—ForDisc, mDNA, a clay facial reconstruction, an experimental computerized reconstruction, and a video superimposition—we were coming full circle, ending up right back where we began, with uncertainty and ambiguity.

I scheduled a meeting for June 1, 2007, to tell Leoma's relatives we'd come up empty-handed. Then, the night before, I received some astonishing news from David Ray, the original TBI investigator. David had long since traded his TBI badge for a sheriff's badge, but he had kept a file on the Patterson case, and he'd just dug that file out of storage. There, to his amazement and mine, he found the hair mat and some bits of dried scalp tissue recovered at the death scene in 1979. Unlike the bones, the hair and scalp had not been simmered in detergent and bleach. We were back in the game.

The following afternoon I briefed the family on all we'd done, and how everything had fallen just short of proving the bones to be Leoma's. They appreciated the efforts, but they were understandably disappointed. When I showed them the hair and scalp, though—which David Ray had brought me that morning—their hopes soared.

I slit the TBI evidence seal on the bag containing the hair and scalp. Snipping off a hank of hair, I sealed it in a ziplock plastic bag, along with a bit of dried scalp, then overnighted the bag to Trace Genetics. Jason planned to divide the samples; he would analyze one half, he said, while his senior technician—working independently in a separate lab—analyzed the other.

On June 18, Jason reported that both he and his technician had found DNA in the hair and scalp. "What we don't know yet is whether it matches the granddaughter's," he said.

On June 26 he knew: The mDNA in the hair and scalp and the mDNA in Michelle's cheek swab were identical. What's more, it was an unusual variety of mDNA, distinguished by two mutations. The odds that the match was random, said Jason, were 1 in 15,000; in other words, the odds that Michelle was descended from the dead woman were virtually 100 percent. After nearly two years and a wild forensic roller-coaster ride, we had identified Leoma Patterson. Conclusively. Positively. Maybe now Leoma—and her family—could finally rest in peace.

In the end, new technology gave us the certainty we sought—technology, plus a little luck and a lot of work by many people: Leoma's family, investigators and prosecutors, dedicated scientists, a gifted artist, and smart, helpful anthropology colleagues. Teamwork remains the key to solving mysteries. No matter how sophisticated the science gets, the people will always, *always* be crucial.

AFTERWORD: WHAT NEXT?

I hope the cases in this book have given you a better understanding of what forensic anthropology is, and how it provides scientific tools that can help solve crimes. I hope these stories have also shown how the research at the Body Farm has steadily sharpened those tools, by providing groundbreaking data on the sequence and timing of human decomposition at the Body Farm; by building a large and ever-expanding Forensic Data Bank, which helps determine the racial or ethnic origin of an unknown victim by allowing comparison with measurements from thousands of other, known skeletons from around the world; by creating and refining ForDisc, a sophisticated computer program that automates the tedious, time-consuming comparison of skeletal measurements; and by producing a steady stream of superb scientists—the backbone, you might say, of forensic anthropology.

For years I've been saying I'm retired, but for years I keep noticing how full my appointment calendar gets. I can't seem to resist invitations to lecture to law enforcement groups and medical professionals and students. I can't turn down an interesting forensic case. I can't stop telling the stories of the dead, and can't stop gathering new ones. Judging by the weight of evidence, I don't want to stop. And why would I? If I'd stopped taking cases when I retired, I'd have missed the chance to X-ray the Big Bopper, and my friend and colleague Jim Starrs wouldn't have called to ask if I'd like to be part of the forensic team slated to exhume and examine the remains of Harry Houdini.

In the grand scheme of things, of course, those cases are interesting, but they're not of earthshaking importance. Meanwhile, though, as I pass along my stories and consult on cases I find interesting, the big, wide forensic world continues to revolve beneath us all. That revolution, I'm proud to say, has been nudged along, and will continue to be nudged along well after I'm gone, by the research and the scientists emerging from a small, smelly, but forensically fertile patch of East Tennessee woods known as the Body Farm.

ACKNOWLEDGMENTS

Like a good forensic case, a good book requires teamwork. As a writing team, we're grateful—both of us—for the continuing pleasure of coauthoring yet another book, our fourth together so far.

Along the way, we've received enormous assistance from many people within the Anthropology Department at the University of Tennessee. Dr. Richard Jantz, head of the Forensic Anthropology Center, has responded graciously and swiftly to numerous inquiries, especially about ForDisc and the Forensic Data Bank. Dr. Lee Jantz, who oversees the osteology lab and the body donation program, has good-naturedly endured numerous interruptions of her own work and that of her graduate students; Lee also helped us conduct what we believe to be the world's first controlled experiment to compare how a human cadaver's underwater sonar signature compares with the sonar signatures of training mannequins.

Dr. Graciela Cabana, UT's new molecular anthropologist, helped us understand more about the subtleties—and the frustrating real-world limits—of mitochondrial DNA testing. Graduate assistant Rebecca Wilson provided valuable help with the sonar experiment, and Ph.D. student Kate Spradley—now Dr. Kate Spradley—boldly took 3-D digitizing where no digitizer had gone before, to a remote cemetery high on a mountainside. Forensic artist Joanna Hughes swiftly created a remarkably lifelike facial reconstruction on the skull we first exhumed some two years ago, and graduate student Megan Moore (along with radiology technician Todd Malone) captured that skull's contours in three dimensions with a CT scan. The Anthropology Department's heroic secretary, Donna Griffin, has done countless favors for us.

Others elsewhere at UT have also been kind and gracious. Among them are Chancellor Loren Crabtree and Vice President Linda Davidson, who have gone out of their way to be supportive and encouraging, and UT Bookstore staff member Susan Seals, who has given tirelessly of her own time (and her husband Jim's time) to help make book signings run like clockwork.

For helping us put sidescan sonar's forensic capabilities to the test, we offer a sincere thanks to the members of the Blount County Sheriff's Office dive team (especially Captain Jeff Burchfield), the Blount County Rescue Squad, the Knoxville Volunteer Emergency Rescue Squad, and the Loudoun County Rescue Squad. For allowing us extended use of a Humminbird 987c sidescan sonar—the little sonar that could—we thank Mark Gibson and Dave Betts of Humminbird.

We're grateful to FBI Special Agents Gary Kidder and Don Provonsha for making it possible for us to share many details about the Roy Lee Clark chop shop and murder case. Also at the FBI, we appreciate Diana Moyers and Phil Williams of the FBI Laboratory in Quantico, Virginia, for helping us try to pin down the elusive face, and the elusive identity, in the Leoma Patterson case.

For courage in the face of uncertainty and pain—and for allowing us to share their stories with the world—we thank Jay Richardson, whose uncertainty ended with the exhumation of the Big Bopper; and we thank the children and grandchildren of Leoma Patterson for their persistence and patience during all the twists and turns of that frustrating and fascinating case. For his heroic efforts to coax bits of DNA from teeth and bones where, alas, there appeared to be none, we express our gratitude to Dr. Jason Eshleman of Trace Genetics, Inc.

Our colleagues at William Morrow seem like extended family to us by now. We would be ungrateful wretches if we did not lavish particular praise on our editor, Sarah Durand; our publisher, Lisa Gallagher; and the publicity and sales teams who turned our first two William Morrow books into bestsellers, particularly Debbie Stier, Kevin Callahan, Ben Bruton, Buzzy Porter, and Emily Krump. Special thanks to Kim Lewis and Andrea Molitor. We also remain thankful to our agent, Giles Anderson, for bringing us into the William Morrow/HarperCollins family.

Finally, we owe our personal thanks to several splendid women. Sylvia Eggleston-Wehr provided a beautiful writing retreat on the banks of the Potomac River, without which this manuscript might never have been finished. Coleen Baird, Queen of the Universe, provided clarity, common sense, and some lovely diversions—also crucial to the writing process. And Carol Bass has surely earned a huge cluster of stars in her crown for the grace and good humor with which she accepts all the disruptions, demands, and outrageous dinner-table conversations that characterize this writing partnership.

GLOSSARY OF KEY ANTHROPOLOGY AND FORENSIC TERMS

Accumulated-degree-day (ADD): the cumulative total of the average daily temperature, measured in either Fahrenheit or Celsius degrees; linking decompositional stages or insect development to ADDs allows temperature changes to be taken into account when computing time since death

Acetabulum: the "socket" in the hip within which the femoral head moves

Adipocere: literally "grave wax," a greasy or soapy substance formed when fatty tissue decomposes in a moist environment

Antemortem: before death

Anterior: toward the front (of the body)

Auricular surface: the surface of the hip bone in the area of the sacroiliac joint

Autolysis: literally "self-digestion"—the breakdown of the body's soft tissues as a result of intracellular chemical changes

Autopsy: a postmortem examination by a forensic pathologist

Blowfly: any of several iridescent green or blue flies in the family Calliphoridae that colonize recently deceased bodies, laying eggs in orifices or wounds; the eggs hatch into maggots that feed on the soft tissues

Cervical: in the region of the neck

Clavicle: collarbone

Coronal suture: joint in the cranium running across the top of the head from one side (parietal) to the other

Coroner: an official who investigates and certifies deaths; a coroner might or might not have medical training

Cranial vault: the top, or dome, of the skull

Cranium: skull

Cremains: human remains that have been cremated

Cremation: incineration of a body

Decomposition: the decay or disintegration of the body

Dentition: the number, pattern, and arrangement of a person's teeth—unique for every person

Distal: far; in bones, farther from the center of the body (the opposite of *proximal*), as in "the distal end of the femur"

DNA: abbreviation for *d*eoxyribo*n*ucleic *a*cid, the long molecule that encodes each individual's complete genetic structure within every cell's nucleus. Shaped like a long, twisted ladder with three billion "rungs"—each rung consisting of a pair of chemical bases—nuclear DNA incorporates genes from an individual's mother and father; however, slight variations in its three billion pairs of bases make every person's DNA unique. As a result, nuclear DNA samples extracted from blood, saliva, semen, skin scrapings, hair follicles, teeth, or bone can serve as a genetic "fingerprint" to confirm a victim's identity, link a suspect to a crime scene or a victim, or (most often in rape cases) clear an innocent suspect. Nuclear DNA—also called the human genome—is far more complex and individualized than *mitochondrial DNA*.

Electron microsope: a powerful, high-resolution variation on the traditional optical microscope. Instead of illuminating samples with light and magnifying the image with optical lenses, an electron microscope "lights" specimens with beams of electrons. Because electrons are so small—far smaller than the wavelengths of light—the images they produce can be magnified by factors of thousands or even millions. A scanning electron microscope is particularly good for creating crisp images of three-dimensional samples.

Epiphysis (plural: epiphyses): a part (parts) of a bone—usually the end—separated from the central portion or shaft by cartilage; specific epiphyses *ossify* at consistent, predictable times, making them important markers of skeletal development or age

Erupt: to break through (the gums, in dentistry). Used often in reference to the third molars ("wisdom teeth"), which, in humans over the centuries, have become progressively less likely to erupt

External occipital protuberance: the bony bump at the base of the occipital bone of the skull, usually prominent in males but not in females.

Femur (plural: femora): the thigh bone (bones)

Femoral head: the "ball" at the proximal end of the femur

Fibula: the smaller, lateral bone of the lower leg

Floater: a body found decomposing in water

Fluoroscopy: the use of ultraviolet ("black") light to excite molecules, causing them to glow with a distinctive and characteristic color

Foramen: an opening or hole (in a bone)

Foramen magnum: the large opening at the bottom of the occipital bone through which the brain stem and spinal cord emerge

ForDisc: short for *f*orensic *d*iscriminant functions, a software program developed at the University of Tennessee by Richard Jantz and Steve Ousley. ForDisc allows forensic anthropologists to enter key measurements from an unknown individual's bones and—by comparing those measurements to thousands of similar measurements from known individuals in the *Forensic Anthropology Data Bank*—the software predicts the sex, stature, and racial, ethnic, or geographic origin of the unknown person. First used in a forensic case in 1991, the program is now in its third version, ForDisc 3.0.

Forensic Anthropology Data Bank (also Forensic Data Bank): a large and steadily expanding set of measurements at the University of Tennessee, taken from well over two thousand modern skeletons from around the world, the majority of them being known, identified individuals. Created in 1986 with funding from the National Institutes of Justice, the Forensic Data Bank provides the comparative foundation on which ForDisc is able to build its calculations of an unknown skeleton's sex, stature, and racial, ethnic, or geographic origin.

Frontal: the bone forming the forehead and upper edges of the eye orbits

Greater trochanter: the larger, lateral epiphysis just below the head of the femur, easily felt as a prominent, bony bump at the top of the outer thigh

Humerus (plural: humerii): the bone (bones) of the upper arm

Hyoid: small, U-shaped bone at the front of the neck, often broken in cases of strangulation

Ilium: broad, upper portion of the hip bone, or innominate

Innominate: hip bone, formed by the fusion of the ilium, ischium, and pubic bone

Instar: any of the three developmental stages of the maggot (first instar, second instar, third instar), distinguished from one another by specific anatomical features, and useful in helping pinpoint time since death

Lateral: toward the side (of the body); the opposite of *medial*

Lesser trochanter: the smaller, medial epiphysis just below the head of the femur

Maggot: the pale, caterpillar-like larva of a fly

Mandible: the lower jawbone

Maxilla: the upper jawbone

Medial: toward the center (of the body); the opposite of *lateral*

Medical examiner: a physician who works with law enforcement officers to determine cause of death

Mitochondria: tiny structures within cells that serve as cellular "power plants" by oxidizing chemicals to release energy. Thought to have evolved from bacteria that cells incorporated many millions of years ago, mitochondria contain their own smaller version of DNA.

Mitochondrial DNA (mDNA): a shorter, simpler, and more plentiful strand of genetic material. It differs from nuclear DNA in several key respects: it is a far smaller, simpler molecule, circular in structure, consisting of 16,569 base pairs, rather than the three billion of nuclear DNA; mDNA is not limited to one copy per cell nucleus, but is found in hundreds or even thousands of *mitochondria* in every cell; mDNA is inherited entirely from one's mother, rather than from both parents; and mDNA changes, or mutates, very slowly, with the result that many generations of mothers can pass along the same mDNA to their offspring. In forensic applications, therefore, mDNA is generally more durable and more plentiful than nuclear DNA, but it cannot serve as a unique genetic fingerprint.

Occipital: the bone forming the back and base of the skull

Odontology: forensic dentistry, most often the comparison of dental records or X-rays with teeth, roots, jaws, fillings, bite marks, and other dental characteristics to identify a body or a suspect

Orbit: the bony socket that cradles the eyeball

Ossify: to turn to bone. At birth, the skeleton is formed of cartilage; as calcium and other minerals reinforce it, the cartilage gradually ossifies.

Osteology: the study of bones

Parietal: literally "of a wall"; the bone forming either side of the skull

Pelvis: literally "basin"; the pelvis is the structure formed by the innominates and the sacrum

Perimortem: at or around the time of death

Phalanges: bones of the fingers and toes

Postcranial: below the cranium, generally referring to the postcranial skeleton (that is, everything from the neck down)

Posterior: toward the rear (of the body)

Postmortem: after death

Proximal: near; in bones, close to the center of the body (the opposite of *distal*), as in "the proximal end of the femur"

Pubic symphysis: the junction at the midline of the pelvis where the left and right pubic bones meet; the features of the pubic symphysis reveal much about skeletal age

Pupa (plural: pupae): insect (insects) in transition from larval stage to adult stage

Puparia: hard, cocoon-like shells in which insect larvae mature into adults; blowfly puparia are often found by the thousands on or near decomposed bodies or skeletons

Putrefaction: decomposition of the body's soft tissues, especially by bacteria

Radiograph: an image produced by *X-rays*

Radius: the lateral (thumb-side) bone of the forearm

Sacroiliac crest: the seam in the hip bone—normally wide, raised, and prominent in adult females—where the *sacrum* is joined to the *ilium*

Sacrum: literally "holy bone"; a triangular bone formed by the fusion of five sacral vertebrae, the sacrum is the posterior part of the pelvis

Scapula: the shoulder blade

Sciatic notch: a gap in the hip bone through which the sciatic nerve passes when it emerges from the lower spine; wider in females than in males.

Sinus: any of four sets of air-filled cavities in the bones of the face and skull. The original function of the sinuses is not clear; however, each person's sinus cavities are unique, and the frontal (forehead) sinus cavity has been shown to be a reliable basis for identification if a portion of the sinus can be matched to an X-ray image of the same area.

Sonar, sidescan: a technology that relies on ultrasound energy to reveal the location and shape of underwater geography and structures. A large-scale cousin of the medical imaging technology that's used to show babies developing within their mothers' wombs, sidescan sonar transmits sound waves and then interprets the acoustic reflections, or echoes, converting them into images of submerged ships, airplanes, automobiles, or even bodies.

Sphenoid: a U-shaped bone making up the middle floor of the skull

Sternum: the breastbone

Suture: here, any of several joints in the skull

Temporal: the bone surrounding the ear

Thoracic: in the region of the chest

Tibia: the larger, medial bone of the lower leg—the "shinbone"

Time since death (TSD): the postmortem interval between death and discovery of a body

Ulna: the medial bone of the forearm—the one that includes the sharp bump of the elbow

Vertebra (plural: vertebrae): bone (bones) of the spinal column

X-ray: a powerful, short-wavelength form of electromagnetic radiation capable of penetrating most solid objects and producing an image, or *radiograph*, of their interior. The denser the object, the less transparent to X-rays it will be; in forensic applications, soft tissue is relatively transparent to X-rays; bone blocks more of the rays; and lead—for example, from bullets—is considered radiographically opaque.

Zygomatic: the cheekbone